Environmental Science and Engineering

Series Editors: R. Allan • U. Förstner • W. Salomons

For further volumes:
http://www.springer.com/series/7487

Arie S. Issar (Ed.)

Progressive Development

To Mitigate the Negative Impact of Global Warming on the Semi-arid Regions

Springer

Editor
Arie S. Issar
Ben Gurion University of the Negev
J. Blaustein Institutes for Desert Research
Sede Boker Campus 84990, Israel
issar@bgu.ac.il

ISSN 1863-5520
ISBN 978-3-642-10639-2 e-ISBN 978-3-642-10640-8
DOI 10.1007/978-3-642-10640-8
Springer Heidelberg Dordrecht London New York

Library of Congress Control Number: 2009943739

Printed on acid-free paper

Springer is part of Springer Science+Business Media (www.springer.com)

This book is dedicated to the memory of Professor Hugues Faure (1928–2003), a pioneer in many fields of science, a friend and a partner in developing the idea of utilization of fossil water found under the deserts.

Hugues Faure was among the first scientists to investigate the impact of Quaternary climate changes on the ground water resources of the Sahara. In this connection, he was the first to discover and investigate the paleo-lakes of the Sahara. In 1968, he was appointed Professor at the Pierre and Marie Curie University of Paris. In 1969, he served as President of the Technical Committee of Geology at ORSTOM ("Office de la Recherche Scientifique d'Outre-Mer") and in 1970 he became Director of the Laboratory for Quaternary Geology (LGQ). From 1982 to 1987, he served as President of the International Union for Quaternary Research (INQUA). All these activities in the cause of national and international science organizations did not impede his creative and incessant research and teaching in the field of Quaternary geology with special emphasize on the study of the carbon cycle during the Last Glacial Maximum (LGM) and the Holocene Epoch.

Let his memory be blessed

Contents

Part II Regional Investigations

List of Figures

List of Tables

Contributors

Eilon Adar Zuckerberg Institute for Water Research, Blaustein Institute for Desert Research, Ben-Gurion University of the Negev, Sede Boker Campus 84990, Israel, eilon@bgu.ac.il

Menakhem Ben-Yami Fisheries Adviser, 18 A.Penn St., 69641 Tel Aviv, Israel, benyami@actcom.net.il

Hendrike Helbron Department of Environmental Resources, UNESCO-IHE, PO Box 3015, 2601 DA Delft, The Netherlands, hendrike22@yahoo.com

Arie S. Issar Ben Gurion University of the Negev, J. Blaustein Institutes for Desert Research, Sede Boker Campus 84990, Israel, issar@bgu.ac.il

Bernhard Lucke c/o Prof. Dr. Rupert Baeumler, FAU Erlangen-Nuernberg, Institute of Geography, Kochstr. 4/4, 91054 Erlangen, Germany, lucke@web.de

Amir Neori Israel Oceanographic & Limnological Research Ltd., National Center for Mariculture, Eilat 88112, Israel, aneori@gmail.com

Iourii Nikolskii Colegio de Postgraduados, Campus Montecillo, Mexico, nikolski@colpos.mx

Dmytro Palekhov BTU Cottbus, Department of Environmental Planning, PO Box 101344, 03013 Cottbus, Germany, dmitry_pal@mail.ru

Part I
Progressive Development – Human Society Can Mitigate the Negative Impacts of Aridization Due to Global Warming

General Introduction

This book joins the list of books authored, edited and co-authored by Arie S. Issar. These books, in addition to the articles by the same author, reported the investigations carried out on the impact of climate changes on the hydrological cycle of the semi-arid regions of our globe. The general conclusions from all these investigations were that during the history of these regions, periods of warm climates had a negative impact on their hydrological cycle. The principle, "The Past is the Key to the Future", prompted the series of investigations, which concentrated on the detailed study of the Circum-Mediterranean region focusing on Middle East.

Chapter 1
The Past is the Key to the Future

Arie S. Issar

As mentioned above the guiding principle in the author's research was that "The Past is the Key to Forecasting the Future". In many respects this reverses the direction, along the dimension of time, of the basic principle of the science of geology, according to which the author was educated. This corner-stone principle, laid down by the founders of the science of geology during the nineteenth century, stated that "The present is a key to the past". This allowed learning from such present geological processes as the mechanisms and duration of erosion and deposition, to be applied to the time scale and processes of past geological eras. Reversing the direction of this principle enabled deduction from past impacts of climate changes during the Holocene (the last 10,000 years) on the hydrological cycle, to what may happen in the future.

The need to adopt this new reversed geological principle became apparent in the mid-1980s, in the framework of the above mentioned International Hydrologic Program (IHP – IV Project H 2-1) of UNESCO and WMO. This project involved an investigation of the possible impacts of climate variations on water management and related socio-economic systems. The investigation of past impacts of climate changes that followed was based on the interpretation of time series of proxy-data, which included ranges of advance and retreat of glaciers, changes in levels of oceans, lakes and rivers, thickness of tree rings and changes in the assemblages of pollen and fauna, as well as the geological characteristics of deposits in aquatic and continental environments. These data together with that of time series of environmental stable isotopes (like $^{18}O/^{16}O$, ^{2}H (deuterium)/^{1}H and $^{13}C/^{12}C$) in ice and sea bottom cores, in tree-rings, and in stalagmites, enabled the reconstruction of past climatic scenarios.

The main conclusions of these investigations were that major climate changes occurred during the history of mankind and that during periods of warm climates humidity decreased in most regions bordering the desert belts of the world. These changes of climate had a negative impact on the natural environment and affected

A.S. Issar (✉)
Ben Gurion University of the Negev, J. Blaustein Institutes for Desert Research, Sede Boker Campus 84990, Israel
e-mail: issar@bgu.ac.il

A.S. Issar (ed.), *Progressive Development*, Environmental Science and Engineering,
DOI 10.1007/978-3-642-10640-8_1, © Springer-Verlag Berlin Heidelberg 2010

the history of human societies. The severity of the impacts on human societies varied according to the extremity of the climate changes. Moreover, it was found that the impact of human actions on the environment played a secondary role, attenuating or intensifying the effects of the natural impact [1–4].

Going back to the principle of " The past is the key to forecasting the future" the prediction was that the present crisis of global warming, this time, most probably, wrought by human deeds, will have severe consequences for the Arid and Semi Arid Regions (ASAR) like the Sahel countries as well as the Horn of Africa, the region south of the Kalahari Desert, Eastern Australia, California and the ring of countries around the Mediterranean Sea including the Middle East. It was also predicted, that, as in the past, the reduction in precipitation will endanger the economies of many countries in these regions.

This prediction was confirmed by various recent investigations presented at the 28th Conference of the Intergovernmental Panel on Climate Change (IPCC) in Budapest during April 2008. The Executive Summary stated that "many ASAR (e.g., the Mediterranean basin, western USA, southern Africa and north-eastern Brazil) are particularly exposed to the impacts of climate change and are projected to suffer a decrease of water resources due to climate change" [5]. Rajendra Pachauri, the chief UN climate scientist and the president of IPCC declared that a decline in agriculture production is forecasted. The IPCC experts estimate that millions of Africans could be afflicted by such water problems by 2020, unless action is taken to mitigate climate change, as water issues would be one of the main problems of climate change.

As a matter of fact this process is more than a prediction; it is already a reality in many parts of the world mentioned above, like the countries of the Horn of Africa and the Sahel.

To reiterate, these negative future scenarios and present Third World scenes of starving people caused the author to question the philosophy of "Sustainable Development" when applied to countries susceptible to drying up from global warming. The basic question posed was whether the natural environment can be sustained when the resources diminish and even disappear?

As the answer to this question was in the negative, an alternative program of development, namely "Progressive Development," was formulated, which is presented in this book.

The first tenet, which this paradigm wishes to fulfill, is the sustainability of human life, decreasing the misery of famine and thirst of the populations in poor countries. The second tenet is progress on the economic level, which implies simultaneously raising the standard of living and furthering the dimension of knowledge. Fulfilling these two tenets can be achieved only through a long term plan of development of the local resources, natural and human. This plan ought to be carried out stage by stage according to a preliminary master plan, which will be checked and adjusted every few years.

In most ASAR around the world this progress will involve the development of the groundwater resources, which will lower the groundwater table. This development will force the population to make use of modern methods of well-drilling and water-pumping, and this would force them to advance their methods of agriculture and marketing. For these purposes, the system of education would have to be continuously upgraded, parallel to the investment in projects to control floods and divert streams in order to enrich the groundwater resources. These projects should progress continuously from the small scale earth dams in small rivers to big dams.

In the case of northeast Kenya, as well as for most of the Sahel Belt, the fulfillment of these tenets should begin with development of the shallow groundwater resources in order to supply water for drinking and for irrigation of small plots to supply food for the hungry population, including the nomads. The more advanced stages of development should aim towards the settling of the nomads in order to enable their education, of both genders, but with special emphasize on the girls, in methods of agriculture. In the more advanced stages the deepening of the wells and the introduction of modern pumping and irrigation methods as well as better crop and fertilization methods will be introduced. Further development of the water resources should include the development of groundwater-enrichment systems by flood control structures, like terraces and small dams, and the preparation and execution of a central controlled system of water supply as well as drip irrigation and fertigation,[1] thereby progressing towards a modern system of agriculture.

Applying this program is rather urgent as recent climatologic data reveal that the current global warming process is gaining momentum, accelerating towards peaks, which overshadow past known historical records. As populations in the danger zones have multiplied, the impact of the global warming on human societies will surpass that of the past, and drought and flooding will exact a very heavy toll. As mentioned, unfortunately, the conventional policy of "Sustainable Development" is not sufficient to avert or even mitigate this global disaster. Thus in order to save millions of people from starvation profound and sweeping changes in the natural environment have to be effected at the expense of as yet undeveloped resources of desert lands and water reserves. The survival and well-being, of future generations of the Third World need to be addressed, giving priority to investment in the planning and development, of new environments while encouraging the local populations in the realm of education and changing their traditional way of life.

References

1. A. S. Issar, Climate change and the history of the Middle East. *American Scientist*, July–August, 83, 350–355 (1995a).
2. A. S. Issar, The driving force behind the cold climate spells during the Holocene, In A. Kotarba (ed.), *Holocene and Late Vistulian Paleogeography and Paleohydrology*, Polska Akademia Nauk, Prace Geograficzne Nr. 189. pp. 291–297 (2003).
3. A. S. Issar and M. Zohar, *Climate Change – Environment and Civilization in the Middle East*, Springer, Berlin Heidelberg, New York (2004).
4. A. S. Issar and M. Zohar, *Climate Change – Environment and History in the Middle East*, 2nd edition, Springer, Berlin Heidelberg, New York (2007).
5. IPCC (2008) http://ipcc.dfinet.ch/pdf/meetings-2009

[1] Advanced method of computerized fertilization by the drip irrigation system.

Chapter 2
Present Global Warming, What Will Be Its Future Impact?

Arie S. Issar

The negative impact of the present global climate change was strongly emphasized by Sir Nicholas Stern, government economic advisor in the United Kingdom. and formerly Chief Economist at the World Bank, His report published on October 31, 2006, described climate change as the "greatest and widest-ranging market failure ever seen", but said that action could and should be taken to avert the worst effects. These effects would transform the physical geography of the world with many millions, of people facing starvation, water shortages or homelessness. There will probably be both more droughts and more floods. An increased incidence of devastating storms is expected. And there is an increased risk of famine in the poorest countries. For reducing the risks of climate change a collective action, namely co-operation between countries is required. Delay would be costly and dangerous. Also the National Climatic Data Center (NCDC) predicts ever harsher droughts, floods, heat waves, and tropical storms as the atmosphere continues to warm.

The Intergovernmental Panel on Climate Change (IPCC) released on Friday 2nd February 2007, entitled "The Physical Science Basis: a Summary for Policymakers". This report concludes that global warming is accelerating due to human activity, namely the ever increasing production of greenhouse gases. These include carbon dioxide, which is the most important greenhouse gas, (from about 280 ppm[1] in pre-industrial period to 379 ppm in 2005). The increase accelerated during the decade prior to 2005 and reached a record on a geological scale of 650,000 years as revealed by the investigation of CO_2 in the water of the cores of the permanent ice caps around the poles. The report points out the correlation between the rise in the level of CO_2 in the atmosphere and the rise of temperature and that eleven of the last 12 years (1996–2006) rank among the 12 warmest years since 1850. The warming trend over the last 50 years is nearly twice that for the last 100 years. Observations since 1961 show that the average temperature of the oceans

A.S. Issar (✉)
Ben Gurion University of the Negev, J. Blaustein Institutes for Desert Research, Sede Boker Campus 84990, Israel
e-mail: issar@bgu.ac.il

[1] Parts per million.

A.S. Issar (ed.), *Progressive Development*, Environmental Science and Engineering,
DOI 10.1007/978-3-642-10640-8_2, © Springer-Verlag Berlin Heidelberg 2010

has increased, which caused a decrease in glaciers and ice caps, which in turn caused the rise of the ocean's level. The total rise for the twentieth century is estimated to be 17 cm. Long term (from 1900 to 2005) trend of decreased precipitation, has been observed in the semi-arid regions of the Sahel and the Mediterranean-type climate of the Mediterranean region and southern Africa. Linked with the higher temperatures, more intense and longer droughts, particularly in the tropics and subtropics, have been observed. Computerized models predict that due to the slow reaction of the oceans, even if the emission of greenhouse gases and other contributors to the warming process were held constant at the level of the year 2000, warming would still continue during the next two decades alt a rate of 0.1°C per decade. Decrease in precipitation, as much as 20% is likely in most subtropical regions. In general mid-range scenarios predict severe droughts and floods, more intense hurricanes and cyclones, pressure on fresh water and food supplies, increased spread of diseases, and rising sea levels that could displace hundreds of millions of people worldwide.

Also according to the records maintained by the World Meteorological Organization (WMO)[2] the year 2006 was estimated to be the sixth warmest year since record keeping began in 1880. In Brazil, heat waves were registered from January to March, including a temperature of 44.6°C (112.3°F) in Bom Jesus on January 31, one of the highest temperatures ever recorded in Brazil. Drought in the south of Brazil caused significant damage to agriculture in the early part of the year with losses of about 11% estimated for the soybean crop yield alone. Severe drought conditions also affected China. Millions of hectares of crops were damaged in Sichuan province during summer and in eastern China in autumn. Significant economic losses as well as severe shortages in drinking water were other consequences. Long-term drought continued in parts of the Greater Horn of Africa including parts of Burundi, Djibouti, Eritrea, Ethiopia, Kenya, Somalia, and the United Republic of Tanzania. At least 11 million people were affected by food shortages; Somalia was hit by the worst drought in a decade. Following the long period of drought the Horn of Africa was hit with severe flooding in 2006, reported to be the region's worst in 50 years. The severity of the floods was increased due to the dry ground which was unable to soak up large amounts of rainfall. The hardest hit areas were Ethiopia, where about 600 people were killed, Kenya where about 723,000 people were affected by the devastating floods and Somalia where some places have received more than six times their average monthly rainfall. In Bilma, Niger, the highest rainfall since 1923 affected nearly 50,000 people. In the same month, the most extensive rainfall in 50 years brought significant agricultural losses to the region of Zinder, Niger. Rare heavy rainfall occurred also in the Sahara Desert region of Tindouf, which produced severe flooding in February damaging 70% of food stocks and displacing 60,000 people.

If these catastrophes were not enough, the coming changes will occur 50% faster than previously thought, say researchers at the Hadley Center, the UK's leading climate-research agency. This will be caused by the dramatic escalation in human's

[2]Environment News Service (ENS), January 3, 2007.

society use of coal and oil, which took place since the late nineteenth century. Thus the present atmospheric carbon level is about 360 parts per million.[3] For the last 10,000 years, the natural carbon sinks, namely the oceans and forests, have maintained atmospheric carbon levels of about 280 parts per million.

Guided by the principle of "the past is a key to forecast the future," the present author concluded that global warming would cause various regions to dry up. Yet, what could not be learnt from the past is the growth rate of the populations in the Third World. Indeed, it seems that the first blessing bestowed upon human beings, according to the Bible, which was to "be fruitful and multiply" (Genesis 1:28) did not take into account that, at a certain future stage of human history, modern medicine and hygiene would reduce infant mortality to such an extent that multiplication in many parts of the world would become a curse. This is especially true in the developing world where populations are growing rapidly. Thus, the developing countries are projected to have 8.2 billion people by 2050. Today, nearly half of the world population lives in cities where, on the one hand, such concentrations facilitate disease transmission, but on the other, they enable the provision of health care. Urban areas are expected to absorb almost all of the population growth in the next 30 years. These people need to be fed, yet from 1950 to 2000 the world cropland area per person shrank from 0.23 to 0.11 ha [1, 2]. Not only is the cropland shrinking; the quantity of water for irrigating these lands is also diminishing, due to a total decrease in the availability of water resources. This was restated at the 4th World Water Forum (WWF), convened in March 2006 in Mexico City, where data were presented showing that more than one billion people currently have no access to safe drinking water. Moreover, an estimated 2.7 billion people, or one-third of the world's population, will face major water shortages by 2025. This shortage of water translates to a shortage of food, as Louise Fresco, former UN Food and Agricultural Organization (FAO) member, predicted that a 70% increase in global food production is required to meet the demands of the world's growing population (WWF, 2006). Yet, according to her forecast, agriculture can still keep pace with the world's demographic growth, once enough water is supplied for irrigation. According to her, the food production required to meet the demands of the world's growing population can be accomplished with only a 14% increase in water use. On the other hand, the map prepared by the International Water Management Institute, as input for the World Water Vision, convened in The Hague 2000 (IWMI, 1998), shows that in 2025 most of the southern part of the world will suffer from water scarcity.

While these forecasts based on field surveys give us the global picture, Fred Pearce's book *When the Rivers Run Dry: Water – The Defining Crisis of the Twenty-First Century* [3] contains a detailed account of the various regions where the catastrophe of the drying up of rivers and aquifers, or the deterioration of their quality, is advancing. In Chap. 6 named: "India a Colossal Anarchy" he tells the story of

[3]As can be learnt from the composition of air entrapped in the deep ice cores this level was not reached in the last 420,000 years.

a region where the water of the rivers was all used up and the only potential water resource remaining is groundwater. This resource is now developed with unbelievable lack of local or national government control. In the last decade, more than 20 million farmers have drilled and pumped their private wells causing groundwater levels to drop from near the surface to depths of hundreds of meters. In Chap. 8 entitled "The World's Largest Mass Poisoning", Pearce reports from other regions of this country about children, who are crippled because their communities depend on water poisoned by fluoride or arsenic. The same is true for wide regions in China, Chile, Ethiopia and Uzbekistan.

If Fred Pearce is right and the world population is running out of water resources, the realization of Malthus's predictions is just a matter of time, and the only artificial remedy is strict birth control imposed by the governments. This was the remedy suggested by Malthus if the natural ones, namely famine, war and plague, are not taking their share. In fact, the only country adopting his prescription is Marxist China.[4]

On the other hand, the communist governed state of Kerala in India is an exception from most of the countries in the Developing World, especially Africa with its growing population, posing anew the problem of how to guarantee the food supply for its steadily growing population. As a matter of fact, Kerala is an example for the materialization of Condorcet's solution, namely voluntary self-control of birth rate by an educated population. This and the basic fact that the forecast of Malthus was falsified in the Western World, which was the subject matter of his study demands an analysis which may point to the possibility of extending the "falsification of Malthusian Catastrophe" beyond the borders of the Western World and the other countries that for the time being are sufficiently fed.

In various publications this falsification is explained by the recent "Green Revolution", namely the introduction of chemical fertilizers, high-yielding plant varieties and the use of pumps and sprinklers for irrigation, utilizing surface and groundwater. All these innovations have kept the world fed though its population doubled during the last three decades. This revolution was preceded by the invention of the steam and, later, electric motors, which enabled the lifting of water by motorized pumps. Another innovation was by the chemist Fritz Haber, who found a way to synthesize ammonia. This led to industrial production of fertilizers, which won him the Nobel Prize in 1918.[5]

---

[4]Karl Marx and his collaborator Friedrich Engels believed that the continuous struggle between the capitalists' class and the exploited-workers' class is driving history. Revolution, which will transfer power into the hands of the workers, will enable industrialization as a part of a centrally planned economy. This would ensure that the needs of all people will be met. Marx and Engels rejected Malthus' population theory, as in their view, ideas like his were contrary to the interest of the workers' class and hindered the social revolution they forecasted.

[5]Speaking on the Nobel Prize initiated by the inventor of dynamite, Haber's name is also connected with the hazardous aspects of science as he initiated modern chemical warfare by promoting and organizing the use of chlorine gas by the German army during the First World War. When Hitler came to power in 1933 Haber had to leave Germany because of his Jewish origin. The German army used his invention to produce the Zyklon B gas, in the concentration camps for the extermination

Indeed Fred Pearce's field evidence explains, in some way, the "falsification" of the predictions of the paradigm of Malthus. The gigantic dams, long pipelines and canals, deep drillings, over-pumping of aquifers etc., which Pearce describes so vividly, were executed mainly in order to supply water for irrigated agriculture. This is backed by the reports of the FAO and IWMI that irrigation uses over 70% of the utilized world water resources. Irrigation freed farmers from dependence on the uncertainty of rain-fed agriculture, especially in the semiarid and even arid regions of the Earth.

To defend the claim that technical innovations alone cannot prevent a catastrophe, one can use the case, which Pearce suggests, of "Israeli farmers raising water productivity fivefold in the past 30 years through a mixture of drip irrigation and recycling treated urban wastewater onto their fields" (p. 340). This achievement was due not only to the irrigation system, which is an achievement by itself, but is also a function of the system of government, education and ideological motivation of the people involved in the promotion of agriculture in Israel. Included in this group have been biologists and agronomists, developing new crop varieties with high yields and tolerating inferior quality water, computer scientists, developing "fertigation" systems that drip to the crop exactly the quantity of water and fertilizer it needs under different growth phases, and highly educated farmers, ready to experiment, and sometimes (as in the case of using brackish water for irrigation in the Negev Desert), to venture ahead of the conservative agronomists, who believed too much in what the text books were saying (Plate 8.1). In the background have been the hydro-geologists who explored for aquifers with innovative methods and politicians and lawyers who formulated a water law, which declared water a national property, the exploitation of which is under the Water Commissioner and public committee control. Today Israel turns to desalination of seawater to close the gap between demand and availability.

These breakthroughs and other recent innovations in the various fields of science and technology, especially in the field of agronomy, suggest that human ingenuity may find new technologies to falsify also the present forecast of a global shortage of food and water due to the growth of the world population. The chapter titled "More Crop per Drop" [3, Chap. 37] confronts the issue of food supply by suggesting the principle of advanced irrigation methods. Yet, with all due respect to the innovations suggested by Pearce, including drip-irrigation (the pride of Israel, where this method was developed), these improvements are unlikely to perform the needed miracle of supplying enough water and food to feed the billions born following a geometrical curve of population growth.

This very special case calls for analysis of the theoretical background of the falsification (to date) of the theory of Malthus and the verification of Condorcet's forecast.

---

of Jews during the Second World War, including Haber's family who did not leave Germany in time.

References

1. J. Larsen, *Population Growing by 80 Million Annually*, Earth Policy Institute, Washington, DC (2002).
2. J. Larsen, *Population Growth Leading to Land Hunger*, Earth Policy Institute, Washington, DC (2003).
3. F. Pearce, *When the Rivers Run Dry: Water – The Defining Crisis of the Twenty-First Century*, Beacon Press, Boston, MA (2007).

Chapter 3
Malthusian and Neo-Malthusian Prophecies of Calamity

Arie S. Issar

Prophecies of calamities involving starvation due to deficiency of rains are abundant in the Judeo-Christian heritage, thundered by the prophets of the Bible, from the warning words of Moses the first prophet to that of the last. Yet, famine, due to drought, was prophesized to occur as a result of the wrath of the Almighty, and as a punishment of not following His commands. As time progressed these commands were differently interpreted by the various spiritual leaders, who arose and offered new versions of the original ones. Yet, natural causes of famine in the framework of a scientific-economic theory were not proposed until 1798, when the book "An Essay on the Principle of Population"[1] by Thomas Robert Malthus (1766–1834) was published. Thomas Robert, the son of Daniel Malthus, was born in the county of Surrey, south of London, UK. The father believed that the application of scientific progress to agriculture and industry would inevitably lead society to affluence. His son Robert did not share this optimism although he did not blame droughts as the reason for famine. His not considering this factor can be understood from his background as he was born in a county of green meadows and woodlands and got his academic education at Cambridge University. Teaching in this university, with its tapestry of green lawns, watered all year long by rain or drizzle, he could not imagine that the lack of water would spell doom for humanity. Moreover, born to the class of the gentry, for whom water was not the vital drink, he concluded that the deficiency of food, not water, would spell famine to the lower class of poor workers. Such a misfortune was inevitable as this population was multiplying, in his time in England, exponentially (2, 4, 8, 16, etc.), on the other hand, food production was growing at an arithmetic rate (i.e. 1, 2, 3, 4), as, he believed that it is just a function of enlarging the area under agriculture. Thus in order that a population will not outrun its food supply, Malthus argued that "positive" as well as "preventive

A.S. Issar (✉)
Ben Gurion University of the Negev, J. Blaustein Institutes for Desert Research, Sede Boker Campus 84990, Israel
e-mail: issar@bgu.ac.il

[1]Thomas Malthus, 1798, *An Essay on the Principle of Population*, An Essay on the Principle of Population as it Affects the Future Improvement of Society with Remarks on the Speculations of Mr. Godwin, M. Condorcet, and Other Writers. London, printed for J. Johnson in St. Paul's Church-yard, 1798.

A.S. Issar (ed.), *Progressive Development*, Environmental Science and Engineering, DOI 10.1007/978-3-642-10640-8_3,

checks" must be applied. In the first category of "checks" which raise the death rate he included disease, famine, and war, while the second category, which lower fertility he included birth control (which he called "Vice"), late marriage, and "Moral Restraint". If the preventive checks are not practiced, then the "positive" natural forces, i.e. famine, disease and war – will come into action.

The book of Malthus, according to its author "owes its origin to a conversation with a friend, on the subject of Mr. Godwin's essay on 'Avarice and Profusion' in his Enquirer". William Godwin (1756–1836) can be regarded as the father of philosophical anarchism. His most important political treatise, *An Enquiry Concerning the Principles of Political Justice, and Its Influence on General Virtue and Happiness* was published in London, and despite its length and cost, it became a bestseller. In the essay to which Malthus refers, the principle, fully developed in *Political Justice* is restated namely: "There is no wealth in the world except this, the labor of man. What is misnamed wealth is merely a power invested in certain individuals by the institutions of society, to compel others to labor for their benefit." Godwin's philosophy of exploitation was cited by Marx and Engels who accepted and supported it in their writings. He did not foresee any calamity due to mounting human birth rate. He believed that intellectual pleasures will replace sexual desires.

Malthus devoted a whole chapter in his book (Chap. 10) to challenge Godwin's social theories. In the first place he argued that it is not scientific, or in Malthus' vocabulary "it is not sound philosophy" ("he (Godwin) has not proceeded in his inquiries with the caution that sound philosophy seems to require"). In the second place Malthus maintains that "The great error under which Mr. Godwin labors throughout his whole work is the attributing almost all the vices and misery that are seen in civil society to human institutions". Malthus argues that "Were this really a true state of the case, it would not seem a hopeless task to remove evil completely from the world, and reason seems to be the proper and adequate instrument for effecting so great a purpose".

The main target for Malthus' criticism is Godwin's statement with regard to the forecast of an overcharged population, by maintaining that "Three fourths of the habitable globe is now uncultivated. The parts already cultivated are capable of immeasurable improvement. Myriads of centuries of still increasing population may pass away, and the earth be still found sufficient for the subsistence of its inhabitants" According to Malthus in the conclusion of this chapter: "In the supposition I have made, I have undoubtedly taken the increase of population smaller, and the increase of produce greater, than they really would be. No reason can be assigned why, under the circumstances I have supposed, population should not increase faster than in any known instance. If then we were to take the period of doubling at 15 years, instead of 25 years, and reflect upon the labour necessary to double the produce in so short a time, even if we allow it possible, we may venture to pronounce with certainty that if Mr. Godwin's system of society was established in its utmost perfection, instead of myriads of centuries, not 30 years could elapse before its utter destruction from the simple principle of population."

Unfortunately, Godwin's forecasts with regard to the future of human society failed. Indeed anarchism was not put into practice, as a system of governing, although they were adopted by certain political movements, yet his socialistic views were part of the ideology of the communist regime and ended in an inhuman bureaucratic regime. At present, not too many societies will rush to try the system in which "men must either have their portion of labor assigned them by society at large, and the produce collected into a common stock on the other hand". In this issue Malthus' realistic opposition to Godwin's views was confirmed. The question remains what about Godwin's prediction that human endeavor will suffice to fulfill all needs of humanity in food and commodities? When it comes to the Western World this forecast was justified in many respects.

As can be seen already from the sub-title of Malthus' book, he also did not agree with the ideas of the French mathematician and philosopher Marquis de Condorcet (1743–1794) which are presented in his book "Esquisse d'un Tableau Historique des Progrès de l'Esprit Humain" written in 1795 [1] while hiding to avoid execution by Robespierre's guillotine. Condorcet not only anticipated much of Darwin's idea of evolution but also extended it to the evolution of the intellect and morality of the human being and his organized social moral code. He looked forward to an era when republican governments would be established throughout the world, and called for complete social, legal, and educational equality between the sexes. He forecasted the social security system and that human life in the remote future could be greatly prolonged by improvement in medical sciences. At the same time the intellectual and moral achievements of humans will bring about the abolition of war while population growth will be controlled by the free will of enlightened families. To his conclusion, which upset Malthus' calculations on the rate of swell of the population, as well as his English conservative code of morals, Malthus responds:

> He (Condorcet) then proceeds to remove the difficulty in a manner which I profess not to understand. Having observed, that the ridiculous prejudices of superstition would by that time have ceased to throw over morals a corrupt and degrading austerity, he alludes, either to a promiscuous concubinage, which would prevent breeding, or to something else as unnatural. To remove the difficulty in this way will, surely, in the opinion of most men, be to destroy that virtue and purity of manners, which the advocates of equality, and of the perfectibility of man, profess to be the end and object of their views.

Malthus considered Condorcet's forecasts as mere speculations, while his own as scientific as they were based on observations and mathematical calculation. Yet, according to his own assertion that "a just theory will always be confirmed by experiment", Condorcet's theories have been confirmed by experimenting in the socio-economic field, concluding with the "Welfare Society" of the Western World. The high standard of living and low rate of birth (for good and worse) of this society are in many ways the realization of Condorcet's "utopia". On the other hand Malthus' theory about the shortage of food in the world, not to speak on the Western World, was falsified. Yet, it is interesting to note that the theory of Malthus had a more profound impact on the world of socio-economic sciences, and even philosophy, than that of Condorcet,[2] this albeit the fact that humans dislike prophecies of doom. The reasons for Malthus' forecasts becoming popular may have been

---

[2]A survey of the internet using Google of the term "Thomas Malthus" gives 676,000 references, "Marquis de Condorcet" 415,000, "Malthusian catastrophe" 75,700, "Theory of Marquis de Condorcet" 52,000.

due, in the first place, to the fact that Malthus based his theory on observations and mathematical calculations, while Condorcet, despite being a mathematician, brought forward intuitive arguments. In the second place the impact of Malthusian Catastrophe Theory on the world of science, was most probably connected with that of the storm, which Darwin's (and Alfred Russel Wallace's) theory raised in public opinion. Darwin found, already in 1838, in the Malthusian law of population a key to understand the struggle for existence of animals and plants.

Towards the end of the twentieth century Neo-Malthusian Catastrophe theory became even more fashionable as a subject of investigation and discussion in connection with a series of droughts which had badly affected the Third World. The media reports showing undernourished children and people starving had a strong effect on public opinion. It prompted surveys and reports by the World Bank, UN and other aid agencies which showed that, taking into account the current rates of increase, the world population is projected to double from roughly 6 billion to more than 12 billion in less than 50 years.

A modern-day equivalent to the prophecy of Thomas Malthus is found in the book *The Limits to Growth* by Donella Meadows [2]. In this book the conclusions of the presentations and discussions of the "Club of Rome"[3] were summarized. Meadows forecasts for the long-term global trends in population, economics and the environment were based on the output of the computer "World3" model prepared and run by her MIT team. According to this model about 3.2 billion hectares of land are potentially suitable for agriculture, world wide of which half is already under cultivation. In order to increase the area immense capital inputs are needed and this "achievement" will have a strong negative impact on the environment, which demands ever more far-sighted and careful use of the resources. However, no matter how carefully the resources are controlled, the inevitable result of unrestrained growth is a catastrophe which can only be postponed. Therefore stabilizing the rate of world population growth must be a continuing priority. Meadows was the founder of the Sustainability Institute, combining research in global systems with practical demonstrations of sustainable living and development namely development that "meets the needs of the present generation without compromising the ability of future generations to meet their own needs".

The same trend of thought was adopted by David Pimentel, Professor (Emeritus) of ecology and systems at Cornell University to adopt a Neo-Malthusian forecast arguing that the number of one to two billion humans which are now malnourished, and which is the largest number of hungry humans ever recorded in history will increase to 3 billion [3]. According to Pimentel the shortage of productive cropland combined with decreasing land productivity is, in part, the cause of current food shortages and associated human malnutrition. His forecasts, with regard to malnourishment are based on reports from the Food and Agricultural Organization of

[3] According to its website, the Club of Rome is composed of "scientists, economists, businessmen, international high civil servants, heads of state and former heads of state from all five continents who are convinced that the future of humankind is not determined once and for all and that each human being can contribute to the improvement of our societies."

the United Nations (FAO) as well as other reports which show that the per capita availability of world grains, which make up 80% of the world's food, has been declining for the past 15 years. In his estimation one-third of the world's cropland (1.5 billion ha) has been abandoned during the past 40 years because erosion has made it unproductive, while repairing erosion losses, namely to form 25 mm of agricultural soil, takes 500 years. According to Pimentel world cropland per capita has been declining and is now only 0.27 ha per capita; in China only 0.08 ha now is available. This is only 15% of the 0.5 ha per capita considered minimal for a diverse diet similar to that of the US and Europe.

While Malthus concentrated his forecasts only on the deficiency of food Pimentel's and his associates' forecasts cover also the problem of diminishing water resources, which are critical not only for drinking but also for irrigation as there is a global trend of over-draft of surface and groundwater resources. Malnutrition and lack of clean water bring to about 90% of the diseases occurring in developing countries. This will intensify in the future as the number of people living in urban areas is doubling every 10–20 years, creating major environmental problems.

Pimentel's recommendations for coping with these problems are in the framework of the philosophy of "Sustainable Development". He thus suggests that "Strategies for the future must be based first and foremost on the conservation and careful management of land, water, energy, and biological resources needed for food production. Our stewardship of world resources must change and the basic needs of people must be balanced with those resources that sustain human life ..." [3]. Yet none of these measures will be sufficient to ensure adequate food supplies for future generations unless the growth in the human population is simultaneously curtailed. In his report Water Resources, Agriculture and Environment he maintains the critical need to implement water conservation practices in agriculture and to control water pollution [4, 5]. The same trend is found in the article World Population, Agriculture, and Malnutrition, where he argues that increases in crop yields have not kept pace with increases in population. Per-capita grain production has been falling, worldwide, for 20 years.

As will be argued later by the present author, in David's Pimentel position based on profound arguments, one cannot find the question, but what to do in arid and semi-arid regions where the water resources are diminishing due to global warming, while the population increases. It goes without saying that the conservation of a diminishing resource, like water due to reasons uncontrollable by the local population is as wishful thinking as halting the global warming, or the global population increase.

Professor Tim Dyson of London School of Economics [6] is less pessimistic than David Pimentel. He is cautious not to declare that progress will be so great that hunger will be banished from the globe and the next few decades could see an increase in the total number of hungry people. Yet, Dyson maintains that a major neo-Malthusian crisis due to severe food shortages in the next two or three decades is unlikely to happen. In his opinion the increase in the number of hungry people is not a decree from heaven, so to speak, "but will depend upon future rates of economic growth, income distribution changes and the scale of future demographic growth”. Dyson points at the fact “that many people in the developing world are poor, uneducated, and have relatively little to offer in an increasingly integrated global marketplace”. The negative role of lack of education, mentioned by Dyson as an important, if not decisive factor in the poverty of the malnourished population, points, in the present author’s opinion, to the action, which has to be taken, namely providing education to these poor people in order to help them to be saved from Malthus’ prophecy. In conclusion, Dyson leaves open the question whether world human population, after reaching 9 billion (a record, which he doubts) while taking into account the negative impacts of climate change, (which may have a positive affect in some regions) will manage to live within the global environment while preserving the conditions for food and agricultural production.

References

1. M. J. A. Condorect, *Sketch for a Historical Picture of the Progress of the Human Mind*, trans. June Barraclough, Weidenfeld & Nicolson, London (1955).
2. D. H. Meadows, *Limits to Growth: A Report for the Club of Rome’s Project on the Predicament of Mankind*, New American Library, paperback, Universe Books, hardcover 1972, New York (1977).
3. D. Pimentel, X. Huang, A. Cordova, and M. Pimentel, *Impact of Population Growth on Food Supplies and Environment*, Presented at AAAS Annual Meeting, Baltimore, MD, 9 February (1996).
4. D. Pimentel, B. Berger, D. Filiberto, M. Newton, et al., *Water Resources, Agriculture and the Environment*, Report 04-1, July (2004).
5. D. Pimental and A. Wilson, World population, agriculture, and malnutrition, *World Watch Magazine*, 17 (2004).
6. T. Dyson, World food trends: A neo-Malthusian prospect? *Proceedings of the American Philosophical Society*, 145(4) (2001).

Chapter 4
The Principles of "Sustainable Development" a Result of Neo-Malthusian Conceptual Model

Arie S. Issar

The Neo-Malthusian ideology professed by the late Professor Donella Meadows led her to establish in 1981 the International Network of Resource Information Centers (INRIC), which distributed, on a global scale, the principles of sustainable development. In 1996 she founded the Sustainability Institute in order to form an educational methodology by which to apply these principles and thus reverse damaging trends in the environment, economy, and social systems.

Already in 1972 the international policy of "Sustainable Development" was initiated at the United Nations Conference on Human Environment held in Stockholm. The initiative for this conference came from the regional pollution and acid-rain problems of northern Europe. The problem that the convening developed countries faced was that of the deterioration of the natural environment, due to excessive industrial development. As the level of education and science in the countries producing the polluting agents was already very high, the adopted solution was to demand that this knowledge be channeled into methods of development consistent with conservation of the natural environment. In 1980 "The International Union for Conservation of Nature and Natural Resources" introduced the term "Sustainable Development" in its publication on World Conservation Strategy. In 1983 the term "sustainable development" gained international recognition after the World Commission on Environment and Development (the Brundtland Commission) was established by the United Nations General Assembly Resolution 38/161. One of its goals was "proposing long term environmental strategies for achieving sustainable development".[1]

"Our Common Future," a report published by The World Commission on Environment and Development in 1987, defined sustainable development as development that meets the needs of the present without compromising the ability of future generations to meet their own needs.

A.S. Issar (✉)
Ben Gurion University of the Negev, J. Blaustein Institutes for Desert Research, Sede Boker Campus 84990, Israel
e-mail: issar@bgu.ac.il

[1] UN Documents http://www.un-documents.net/a38r161.htm

A.S. Issar (ed.), *Progressive Development*, Environmental Science and Engineering,
DOI 10.1007/978-3-642-10640-8_4,

In 1992, the issue of sustainable development passed from the developed countries into the sphere of the underdeveloped ones and became international policy. This happened when the United Nations adopted it in the form of the "Rio Declaration on Environment and Development", formulated at The UN Conference on Environment and Development in Rio de Janeiro, and adopted as Agenda 21 by the UN Development Program.[2] It is interesting to note that of the 27 principles expressed in 1,280 words, the term "sustainable development" is mentioned eight times, "poverty" is mentioned only once, and the word "education" is altogether absent.

One would expect that a UN conference on environment and development would set the poverty stricken populations in the underdeveloped countries as its main target and their advancement by helping them to achieve material and educational projects as a first priority goal. It therefore seems as though an issue that was important, if not crucial, for the overdeveloped countries, i.e. "Sustainable Development" was forced upon countries whose immediate problem was survival.

This omission was corrected in the 2002 World Summit on Sustainable Development in Johannesburg. At this summit, the main objectives of sustainable development were defined to be:

(1) Eradicating poverty,
(2) Protecting natural resources,
(3) Changing non-sustainable production and consumption patterns.

It is important to note that the priorities of the UN's projects changed once the majority of participants came from the underdeveloped countries, thus, the "Johannesburg Plan of Implementation" committed the United Nations to a new global partnership to reduce extreme poverty and setting out a series of time-bound targets, all with a deadline of 2015: these have become known as the "Millennium Development Goals" (MDGs).

In this declaration the goals are the following:

(1) Eradicate extreme hunger and poverty,
(2) Achieve universal primary education,
(3) Promote gender equality and empower women,
(4) Reduce child mortality,
(5) Improve maternal health,
(6) Combat HIV/AIDS, malaria and other diseases,
(7) Ensure environmental sustainability,
(8) Develop a global partnership for development.

The fact that environmental sustainability is only number 7 in the list of MDGs after the whole host of problems specified by the various participants at the 2002

[2]United Nations (1992).

summit, who convened under the banner of Sustainable Development, one cannot escape the conclusion that this banner, although of primary importance to the developed world, is of low priority in the developing world.

Since the summit in Johannesburg, the threat of the hazards, which climate change can inflict on developing countries has become rather obvious. Thus, although problems stemming from this change are mentioned in the report of this summit, they are restricted to just a few regions, such as the low-lying islands. Investments to mitigate the negative impacts of this change were not considered important enough to be included in the main objectives of the summit. Since then, Nature has demonstrated the destructive force of the climate change. In East Africa, lack of rains for 3 years has left 2.5 million Kenyans close to starvation, while 1.4 million people in Somalia and 1.5 million in Ethiopia are at risk of starving to death, according to UN estimates. Moreover "The Human Development Report of 2006 tells us that during 2005 more than 20 million people in the Horn of Africa were affected by drought.

A report, released by the UN Development Program (UNDP) in November 2006, finds that even with an agreement to reduce carbon emissions through international cooperation, dangerous climate change is now almost inevitable. The most severe consequences will be experienced by countries and people who bear little responsibility for the problem, it notes. Agriculture will be hit the hardest because of changing rainfall patterns, malnutrition could increase by 15–26% by 2050. Already, lack of water is ravaging many parts of Africa. In northeast Kenya, some 3 million people risk starvation from drought, violent clashes between farmers and pastoralists are increasingly common and the country's gross domestic product fell 16% between 1998 and 2000 due to drought.

Chapter 5
The Theoretical Basis Behind the Falsification of the Theory of Malthus and Verification of Condorcet's Confuted Model

Arie S. Issar

Before discussing the new program of Progressive Development it is important to analyze the wider theoretical basis on which the conceptual model of Malthus was based. As mentioned in arguing against Condorcet and Godwin he claimed that his predictions are based on observations and calculations, while the works which he debated were written as philosophical speculative essays. Malthus' methodology has set the foundations of modern science of economics and no wonder that the his essay, albeit written about 200 years ago, is obligatory reading for every student of economics and is still included in the reference list of modern articles about economics. On the other hand, the forecast of Condorcet, which was written by a leading mathematician, was based on intuitive ideas and ideals, rather than observations and calculations. Yet as already mentioned, although based on the principles of science, the forecast of Malthus has proven false, even when it comes to most countries of the Third World, not to speak of the Western World, which was the focus of Malthus' forecast. At the same time that of Condorcet has been verified. As the present essay is written by a hydro-geologist, who investigates the impacts of climate changes on the global water resources, and thus on food supply, with the aim of using past records in order to forecast future catastrophes, it is important to analyze in more detail the reasons which have verified the forecasts of Condorcet.

Jean Marie Antoine Nicholas Caritat Marquis de Condorcet, who was born in 1743, did not lack the scientific tool of mathematics. When he was just 22 years old he already wrote a treatise on integral calculations. He was the friend of the distinguished mathematician d'Alembert, himself a pioneer in the study of differential equations and their use in physics, and with Diderot composed the monumental 28 volume Encyclopedia of which d'Alembert wrote most of the mathematical articles. Thus although Condorcet did not base his thesis on mathematical calculations, still the principle of mathematics, namely logical reasoning based on axioms, was integrated into Condorcet's social theory. This integration took place by the fundamental argument that by appealing to reason and fact then there is no limit to the

A.S. Issar (✉)
Ben Gurion University of the Negev, J. Blaustein Institutes for Desert Research, Sede Boker Campus 84990, Israel
e-mail: issar@bgu.ac.il

A.S. Issar (ed.), *Progressive Development*, Environmental Science and Engineering,
DOI 10.1007/978-3-642-10640-8_5,

progress and thus achievements of mankind albeit the constraints, which nature may impose. In Condorcet's words: "The aim of the work that I have undertaken, and its results will be to show by appeal to reason and fact that nature has set no term to the perfection of human faculties; that the perfectibility of man is truly indefinite; and that the progress of this perfectibility, from now onwards independent of any power that might wish to halt it, has no other limits than the duration of the globe upon which nature has cast us" [1].

Condorcet put special emphasize on "the progress of knowledge and that of liberty, virtue and respect for the natural rights of man" due to the fact "that man is a sentient being, capable of reasoning and of acquiring moral ideas." His optimism led him to believe that "The time will therefore come when the sun will shine only on free men who know no other master but their reason ... and to learn how to recognize and so to destroy, by force of reason, the first seeds of tyranny and superstition, should they ever dare to reappear among us."[1] These anticipations, based on reason, brought him also to predict the diminishing rate of increase of population, and thus avoid the threat of insufficiency of food. The fact that Condorcet's conceptual model proved correct, and not the methodical and exact predictions of Malthus, points to the fact that the basic principle, namely the power of human reason, which exists in the philosophy of Condorcet and is lacking in that of his opponent, is a crucial factor.

Putting Condorcet's prediction in modern wording it can be said that he predicted that human intelligence will progress in such a way that will enable it to survive against the constraints due to Spencer's-Darwinian paradigm of "survival of the fittest" but according to the paradigm of Jonas Salk "The survival of the wisest" [2]. In other words, Condorcet forecasted the evolution of the "Industrial Revolution" and "Green Revolution", as an obligatory result of the progress of the human mind, namely the evolution of intelligence.

The fact that Darwin considered the theory of Malthus as a major tool to understand the mechanism behind natural selection explains Darwin's negligence of the role of the evolution of intelligence in the bio-world. The interesting fact is that Darwin adopted a survival model based on observations and calculations made on the human society, to explain the mechanism of natural survival in the bio-world. Yet, because the base of Malthus' survival model did not take into consideration the role of intelligence as a survival mechanism, Darwin also did not take it into consideration, when he applied Malthus' model to the bio-world. The fact that Malthusian-Darwinian model did not work when it came to the human race, while Condorcet's model did work, speaks to the role of intelligence in the natural selection mechanism of *Homo sapiens sapiens.* [2]

---

[1] M. J. A. de Condorect, pp. 9–10, 128, 136, 140–142, 173–175.

[2] For further reading on role of intelligence on evolution in general and the bio-world in particular the reader is referred to the following publications: A. S. Issar with Robert G. Colodny [3]. A. S. Issar [4].

To explain the role of intelligence in the evolution of mankind in more general terms it is suggested to regard the progress of man's intelligence as progress along the *dimension of information.* In other words, the author suggests that there exists an additional dimension to space-time along which knowledge and thus intelligence is measured. It is the dimension by which growth of any human being is measured in addition to his age (time) and size (space). Thus, knowledge, based on experience, intelligence and science, as well as the social moral code, are structures on this dimension. This science-philosophy model explains the success of Condorcet's paradigm and the failure of the Malthusian paradigm in the framework of a general "coordinate" system, which includes *information* in addition to space-time. Thus the evolution of the Western World after the Scientific Revolution, starting in the sixteenth century with Nicolaus Copernicus, was launched on the coordinates of the *dimension of information*. It brought in its wake the Industrial Revolution, starting with the invention of the steam engine at the end of the seventeenth century and followed by the invention of the "spinning jenny"[3] in the eighteenth century. The introduction of machines, which are "materialized information", changed the foundation of European economy. When it comes to the specific problem of supply of food, machines revolutionized agricultural methods in all that concerns cultivation, harvesting, processing and transport. In regions dependent on groundwater, the mechanized drilling machine and vertical pump spelled a breakthrough enabling larger quantities to be brought from greater depths.

The answer to the question: What may be the trend in the future? depends whether the Developing World will be able to stop expanding in number and space (space-time dimensions) and start evolving along the *dimension of information*. For this shift a tremendous investment in education has to take place, including all fields of the natural sciences and at all levels from kindergartens to universities.

Fortunately cyber-space, which disseminates the world of knowledge along the *dimension of information*, and personal computers, could serve as the vehicles enabling exploration of the world along this dimension, and may be available at a low cost, thus enabling remote communities to achieve knowledge, once instruction is made available.

Thus, water- and, consequently, food-sufficiency in the world depend on whether the developed countries can supply enough computers and instructors for the developing world to carry out its progress along the Dimension of Information. Such an investment will, in the long run, undoubtedly pay off as it will increase the numbers of consumers of sophisticated products manufactured by the developed countries. Such progress, which has started to gain momentum in various countries, like the Vision 2020 project in Rwanda [5] and the US $100 laptop project [6], depends in the first place on the contributing countries, but no less so on the developing countries already positioned on the brink of famine and which forecast of the two scientists, i.e. Malthus or Condorcet, will be verified?

---

[3] A hand-powered multiple spinning machine that was the first machine to spin more than one ball of yarn or thread.

References

1. M. J. A. de Condorect, *Sketch for a Historical Picture of the Progress of the Human Mind*, trans. June Barraclough, Weidenfeld & Nicolson, London, pp. 4–5 (1955).
2. J. Salk, *The Survival of the Wisest*, Harper and Row, New York/London (1973).
3. A. S. Issar and R. G. Colodny, *From Primeval Chaos to Infinite Intelligence – Information as a Dimension and Entropy as a Field of Force*, Ashgate, Avebury (1995).
4. A. S. Issar, Evolution of intelligence in the Bio-World. In: R. Gordon and J. Seckbach (eds.), *Divine Action and Natural Selection: Questions of Science and Faith in BiologicalEvolution*, World Scientific, Singapore (2008).
5. M. Kakimba, *Rwanda in her Vision*, The New Times Publications Sarl, Kigali–Rwanda, Friday 4 August (2006).
6. M. Ricciuti, *The $100 Laptop Moves Closer to Reality*, CNET News.com, September 28 (2005).

Chapter 6
"Progressive Development" the Modern Version of Condorcet's Conceptual Model

Arie S. Issar

At the root of the conceptual model of "Progressive Development" is Condorcet's claim that the human mind, educated properly, is equipped with the faculty to supply the material and needs of its society. As mentioned above, Malthus did not agree with this basic idea and claimed that Condorcet belonged to "a set of artful and designing knaves who preach up ardent benevolence and draw captivating pictures of a happier state of society" and are not "aware of the tremendous obstacles that threaten, even in theory, to oppose the progress of man towards perfection". The present author's impression is that one of the main reasons for Malthus' harsh words was his English proclivity to conservatism and thus his suspicion and fear that there exists a conspiracy of knaves who bring up unrealistic theories in order "to enable them to destroy the present establishments and to forward their own deep-laid schemes of ambition, or as wild and mad-headed enthusiasts whose silly speculations and absurd paradoxes are not worthy the attention of any reasonable man".[1] As the most reliable scientific test, namely, the "falsification test", proved Malthus to be wrong and Condorcet correct in all that related to European society, to which Malthus was referring, the logical conclusion would be to follow Condorcet's proven theory. This theory foresaw that the advance and diffusion of knowledge would result in the establishment of an enlightened liberal democracy, in which women would have the same rights as men, and birth control would be voluntary. The reduced population growth together with the advancements in agriculture would eliminate the threat of hunger. The rectification of this theory encourages us to extend it to the current problems of our generation and the generations to come. Thus, the first concern of any development plan should be the promotion of knowledge.

A.S. Issar (✉)
Ben Gurion University of the Negev, J. Blaustein Institutes for Desert Research, Sede Boker Campus 84990, Israel
e-mail: issar@bgu.ac.il

[1] T. Malthus, *An Essay on the Principle of Population*, An Essay on the Principle of Population as it Affects the Future Improvement of Society with Remarks on the Speculations of Mr. Godwin, M. Condorcet, and Other Writers. London, printed for J. Johnson in St. Paul's Church-yard, Chap. 1 (1798).

A.S. Issar (ed.), *Progressive Development*, Environmental Science and Engineering,
DOI 10.1007/978-3-642-10640-8_6,

Expressing Condorcet's philosophy in the framework of the present author's aforementioned conceptual model, in which information is a dimension along which knowledge is structured, any development plan which aims at the human or natural environment must precede with progress of the human mind, namely, knowledge. This can be achieved by first investing in educating the population that is destined to benefit from the investment in changes in the natural environment. A society educated in science and technology will ensure that the human mind can find sophisticated new ways to guarantee survival at a higher level of sustainability. As a matter of fact, these have been the landmarks in mankind's evolution: *Homo sapiens sapiens* invented irrigated agriculture when the Near East was becoming more and more arid after the Last Glacial humid period, developed mining and smelting of metals to enable the expansion of populations during times of want in Central Asia due to cold periods, developed pumps to lift water when the springs feeding the aqueducts that had been built all over the Roman-Byzantine Empire dried up due to the Arab warm period [1, 2]. The Industrial Revolution started with the invention of the steam engine to help pump water out of coal mines. That this invention took place during the Little Ice Age, which began around 1650 when coal was the main fuel for warming houses, may not be just a coincidence. The cybernetic world, into whose endless boundaries the human mind is expanding today, is an illustration of expansion along the dimension of information, and is furthering Condorcet's vision of the worldwide spread of enlightened education.

When the present author was involved in the development of water-resource projects in developing countries, during the 1960s and 1970s, he had, from time to time, to teach the local young people to drive a car or a tractor, replacing the traditional horse-driven vehicles. This teaching enabled these people to move faster along space-time dimensions but simultaneously enabled them to progress along the dimension of information in all that concerns motorized machines. As vehicles, are rather common in the Developing Countries, the next step required in the twenty first century in these countries is acquiring the ability to travel in the expanses of the cybernetic world. This travel will enable young people in these countries to gain knowledge, which was once stored in libraries and in the brain cells of learned people.

Progressive Development projects should thus start with teaching the ability to handle a computer and travel on the internet. The vehicles, namely the computers, will hopefully become more available, as the MIT project to develop a $100 laptop computer is starting to gain momentum in various countries [3]. Once this project fulfils its purpose, the next stage should be to enable every young person to have a computer and know how to use it. In parallel, emerging from Condorcet's principle that education is the trigger for development, it is suggested to follow the example put forward by the 2006 Nobel Peace Prize recipient, the Bangladeshi economist Muhammad Yunus and the Grameen Bank he created, but to shift the crediting to the area of education. For this a special international bank, with branches in developing countries, should be founded. This bank will channel loans to young people who wish to advance their education but cannot afford it. The term of the loan will be annual but will be extended according to the yearly certificate and will have to be paid back in annual installments, including interest, after the student has finished his studies and gotten a job. In addition to the supply of computers, teachers knowing how to use these information vehicles will be needed. For this purpose, it is suggested that in parallel to the laptop project, the preparation of instructors be undertaken. For this purpose, university graduates in the developed countries will be specially trained and sent to work in the developing countries. Although unemployment rates for university graduates in developed countries is lower than the average for non-graduates, the problem of employment for university-educated young people still exists and draws from the national budget in order to compensate the unemployed graduates. These budgets should be, in the first place, for special courses to prepare the graduates as teachers in high schools and colleges in developing countries and funds to sustain these teachers in those countries. Such an endeavor will bear fruit for all involved parties.

Progressive Development will not be restricted to the dimension of information. It must be followed by development (not necessarily expansion, although this is permissible) in the dimension of space, which will target the natural resources of the world. This may also include the exploitation of one-time reserves, such as fossil fuels, water resources and other earth resources such as metals, minerals for the production of chemical fertilizers and cement. This statement will undoubtedly raise voices of protest from the zealots of Sustainable Development, whose secure and comfortable way of life is almost wholly a function of the exploitation of one-time non-sustainable reserves. The challenge faced by Progressive Development is the parallel investment in research in order to develop alternative resources, which will replace the exhausted ones. These resources will include new materials, as well as environmentally friendly sources of energy.

Yet, the most important resource in the arid and semi-arid regions is water. A survey of the reports from these regions, in the various media shows that thirst and starvation from lack of this resource will become even more extreme due to the impact of the advancing warming of the globe. "Progressive Development" is based on the lessons, of successes as well as of failures, learnt from various projects in which the author was involved from the dry provinces of South Korea, Iran, Nepal, Brazil, Columbia, Venezuela, Kenya, Republic of South Africa and others.

The principle aim of Progressive Development is to gradually change the natural and human environments, in order to enable survival against the deterioration of climatic conditions, and strive to guarantee progress towards safer and higher standards of life.

The stages for fulfilling these aims are the following:

1. On the basis of a preliminary survey prepare a regional development plan, starting with the exploitation of the most available and low cost water resources and progress stage by stage to more costly resources and advanced methods of exploitation.
2. Construct an infra-structure of an extension service and schooling for the population involved in the project, covering technical as well as cultural fields.

3. Promote demonstration projects by the local inhabitants, backed by an advisory team.
4. Promote the settling of families of the nomad population by offering them a piece of land and a water resource on which they can continue holding their livestock but ensure their water and food supply without moving from one place to another.
5. Move to more advanced and sophisticated stages once basic supply of water and food has been guaranteed.
6. Promote local industry for the supply of equipment, spare parts etc.

As can be seen, the principle of this program of development is to encourage the population, on the threshold of starvation, through a gradual and voluntary process, to change their traditional way of life. This involves, in the first place, teaching the sedentary-pastoral and dry farming population to develop the local shallow groundwater resources and shift to an agricultural way of life, dependent upon irrigation methods. At the same time the program aims towards settling the nomadic-pastoral population in the provinces where the nomadic way of life is the primary or secondary socio-economic system.

It is envisaged that projects following the policy of Progressive Development, which may involve the utilization of one-time reserves like the pumping of groundwater, especially fossil water, will undoubtedly be regarded as non-sustainable development. Indeed this is a valid argument but it does not take into considerations that the natural resources in the arid and semi-arid regions diminish naturally because of the global warming. At the same time the conceptual model of Progressive Development puts the emphasis on innovative activity, that is, the development of new resources by scientific research and development. The philosophy of Progressive Development requires that during the period in which the one-time reserves are being consumed, the economy of the region will be Progressive to a level that would enable the construction of a more sophisticated project to produce a new income resource to replace the consumed one. Such a project, for example, can be the diversion of rivers from more humid regions to desert regions, as in the aforementioned cases. Another example is the desalination of brackish water existing below the freshwater fossil aquifers.

All in all, projects should be planned and carried out as integral components of a comprehensive plan of regional development preceded by research into the hydrology, soil and ecology of the region, and followed up by monitoring and re-assessment procedures. Special emphasis should be placed on monitoring programs as a part of management planning to avoid salination damage to the soil and water resources. An important constraint, which has to be taken into account, arises from the fact that through their roots, most trees extract only fresh water from the soil and leave the salts in the subsurface. This causes the load of salts in the water cycle to increase towards the outflow. Natural flow systems drain out the salt load, either to the sea or to an inland salt marsh or lake where the water evaporates and the salts accumulate. Thus any regional project involving the re-planting of semi-arid and arid zones should be preceded by a regional hydrological and hydro-chemical flow model. The output of such a model would enable decisions on the number of trees per watershed, the rate of increase in salt content, the planting of more salt-tolerant trees downstream, and the location of an outlet area, either natural or artificial, as a sinkhole for the salts.

Another consideration is the fact that planting on permeable rocks may reduce recharge to groundwater. In such cases, deciduous trees which, in cold humid winter regions do not transpire during the cold rainy season, should be planted. This would enable the subsurface water to reach the depth below the root zone during the rainy season.

Here we come to the issue of mitigating the greenhouse effect. One of the aims of Progressive Development is to develop new ways of mitigating global warming, either by developing new sources of energy, or by developing methods of sequestering the greenhouse gases emitted into the atmosphere (by industry and vehicles) by planting vast areas in regions without vegetation, irrigating them by the water resources aforementioned.

When it comes to human socio economic scenarios, the reader, along with TV audiences all over the world, has undoubtedly been exposed to the distressing pictures of starvation, including children, every time a drought strikes the regions bordering the deserts. These sights prompt the international agencies, governments and NGOs in rich countries to organize shipments of food, which also get full media coverage but overlook the fact that this aid measure is temporary. In the pre-television era, when such famines occurred, these people were either left to die or to migrate into the more humid regions, either peacefully, or by force. Such movement was restricted, however, since political borders, demarked by the colonial nations, interfered with the free movement of nomads and herders. It is quite possible that as the global climate changes become more pronounced, the frequency of such droughts, and hence pictures of suffering, will increase. The aforementioned case of northeast Kenya is an example of the disingenuousness of those who opposed the suggested settling of the region's nomads, in order to save them from future starvation and misery.

Progressive Development aims to minimize the suffering of populations in regions that are predisposed to frequent droughts, and hence starvation, due to the global climate change. Any solution has to take into consideration that the present political constraints restrict the range of migration, while at the same time, the processes which in the past controlled the proliferation of the population, namely infant death and starvation, have been somewhat mitigated.

Analyzing this question logically brings us to the conclusion that the problem lies in the fact that the nomadic and semi-nomadic systems evolved for survival while depending upon an accidental supply system. One solution was the storage of food, another was movement. These solutions failed once droughts became too frequent and the borders closed. Thus, once the accidental nature of the system becomes more extreme, the solution lies in stabilizing it, thereby increasing its reliability. As discussed above, in most arid regions groundwater storage exists and can be used to increase the supply of food. The storage of groundwater is possible due to the physics of liquid flow through porous media. The porosity is created by the empty spaces between small rock particles, around which water flows. These particles form natural minute retarding dams, thus reducing the flow velocity. This effect is cumulative and, under special geological conditions, may be responsible for storage over many years. The pumping of groundwater lowers the water table, and in many cases this translates to mining of the resource. Yet, once the preliminary hydro-geological study ensures a supply for a few decades, if not centuries, and the chance for further development of more remote (such as the diversion of a river), or deeper, or any other water resource (a dam for example), the project should proceed, even though it is based on a non-sustainable process.

A plan based on the principles of Progressive Development should move in the direction of changing the traditional socio-economic system from nomadic and herder to settler, through the provision of a stable supply of water, based on shallow or deep groundwater, in parallel with a training program, supply of tools, seeds, plants, and so forth. Such a plan, designed on a regional basis, can be part of the regional and global project of planting these regions, as each settler will be entitled to a plantation of trees most suitable to his region, and to help in getting the plantation started.

References

1. A. S. Issar and M. Zohar, *Climate Change – Environment and Civilization in the Middle East*, Springer, Berlin, Heidelberg, New York (2004).
2. A. S. Issar and M. Zohar, *Climate Change – Environment and History in the Middle East*, 2nd edition, Springer, Berlin, Heidelberg, New York (2007).
3. M. Ricciuti, *The $100 Laptop Moves Closer to Reality*, CNET News.com, September 28 (2005).

Chapter 7
Development of Groundwater, the Fundamental Resource of Projects Based on the Principle of Progressive Development

Arie S. Issar

An important question to be answered is whether there is enough water in the arid and semi-arid regions to enable development projects on a global scale. The answer is positive for extensive portions of these lands [1, 2]. Hydrological and hydro-geological investigations carried out in these regions worldwide have shown that ample water resources are available, and as in the past, whether this water can be utilized to a positive end is simply a question of human ingenuity. Indeed, since the invention of agriculture about 10,000 years ago, farming societies have found ways to make use of the available water resources along the margins of the world's deserts and in many cases inside the deserts. Thus, if the human being could achieve this when technology was rudimentary, he should be able to make use of the existing water resources, even if they become scarcer due to the present climate change. In this case groundwater is the ideal resource, which enables Progressive Development.

The special importance of groundwater for a semi-arid country arises from the fact that this resource always contains a storage factor which makes the water supply less dependent on the random nature of precipitation in such regions. In other words, because the key to maintain a stable income level in agricultural communities is a stable water supply, groundwater storage renders a more deterministic character to the water supply system which otherwise is stochastic and is ruled by random climate changes.

Some of the aquifers under arid and semi-arid lands are recharged yearly, some only once every few years, by precipitation falling on the desert, by infiltration from floods into alluvial fans, or by subsurface flow from adjacent highland areas where precipitation is more abundant. In many areas, this subsurface water emerges at the surface and evaporates to form salt marshes. As this resource is found at various depths and stored quantities, development programs can be applied according to the principles of Progressive Development, namely first on a small scale that requires a low initial investment, yet guarantees the survival of the population even as climatic conditions worsen.

A.S. Issar (✉)
Ben Gurion University of the Negev, J. Blaustein Institutes for Desert Research, Sede Boker Campus 84990, Israel
e-mail: issar@bgu.ac.il

A.S. Issar (ed.), *Progressive Development*, Environmental Science and Engineering,
DOI 10.1007/978-3-642-10640-8_7,

In addition to the aquifers, recharged from time to time, under most of the world's deserts, there also exist large amounts of groundwater which were recharged when climatic conditions were much more favorable, tens of thousands of years ago. This water moves very slowly, either towards springs (like the Mound-Springs in Central Australia), salt marshes (like the Shotts of the Sahara) or towards inland lakes (like the Dead Sea). In Australia, Saudi Arabia, Libya, Jordan and Israel, some of this water is utilized for agricultural and industrial purposes [1]. Needless to say, use of this water means that it has to be mined, which may be regarded as non-sustainable development; however, at the same time, one should also regard this resource as having the potential to ameliorate the ill effects of burning fossil fuels, once it will be used for extensive tree plantations in the deserts (See attachment: UNESCO A Call for Global Action, 1998). Thus, if humanity cannot avoid the use of fossil fuels, and thus cause harm to the environment, why shouldn't a fossil resource, namely paleo-water, be used to correct this harm? The question which always arises when this proposal is brought up is: what happens in the future when this resource is exhausted? Progressive Development planning takes account of this concern and at the early stages of the regional planning, considers future regional projects like that of "Replenishing Lake Chad", which entails damming the Oubangui River at Palambo before it joins the Kongo River in the Central African Republic (CAR) and diverting some of its water to Lake Chad.[1]

Isotopes, radioactive and stable, play an important role in the investigation of the fossil aquifers. The first type disintegrates to another type of element while emitting energy as radiation. Thus carbon 14, which is produced in the upper atmosphere by cosmic rays, disintegrates to nitrogen 14 producing β radiation. The half life of carbon 14, namely the time needed for 100% of the isotope to reach a 50% value is about 5,730 years. This allows dating sediments and water and archaeological artefacts containing carbon to the maximum range of about 40,000 years. The half-life of tritium (hydrogen 3 or ^{3}H) is about 12 years. As a result the maximum dating of water by this isotope is only in the range of 40 years. The water's age can thus be determined either by the tritium content (if the water is less than 40 years old) or by ^{14}C if not older than 40,000 years.

While carbon 14 and tritium are unstable isotopes, there exist stable isotopes, which are used in groundwater research. The stable isotopes Oxygen 18 and hydrogen 2 (deuterium) have been part of the earth's hydrosphere since the beginning. They do not disintegrate and can be distinguished by their heavier weight than oxygen 16 and hydrogen 1. Due to the differences in weight, water molecules containing more ^{2}H and/or ^{18}O will evaporate in smaller quantity than the lighter water composed only of $^{1}H_2{}^{16}O$. Therefore, the relative quantity of heavy isotope in water evaporated from the sea, forming clouds and rain, is less than that of ordinary ocean water. Yet the isotopic composition of the rain will change according to the temperature of the seawater and that of the atmospheric environment when the rain falls. When this happens the heavier rain will fall first and the lighter later. The same is

[1] UN's IRIN (2003).

true with regard to low and high altitudes. Thus when analyzing the isotopic content of the water one can say where the rain fell and infiltrated into the subsurface.

The rainwater reaching the surface of the land will again undergo the process of evaporation. This will further affect the water's isotopic composition, causing it to become heavier in these isotopes. However, the water, which infiltrates immediately to the subsurface to form groundwater, will escape further evaporation. The isotopic composition of water serves as a very good clue to the environmental conditions under which the water was recharged, especially when this information is used in combination with the ordinary chemical analysis of the water, i.e., the quantity and ratio of the salts dissolved in the water.

In addition to the groundwater resources surface water is also available in the arid and semi-arid regions, yet this source varies according to the seasons and may change from 1 year to another. Various investigations of the hydrological regime in arid countries have shown that the percentage of runoff volume relative to the total drainage basin increases relative to the decrease in the size of the drainage basin. A series of investigations carried out during the last two and a half decades in the framework of the J. Blaustein Institutes of Desert Research at the Ben-Gurion University of the Negev, Israel, have shown that about 10–20% of the runoff water is wasted as flash floods, causing erosion and other damage before they reach the sea or saline lakes. In addition, water is wasted due to capillary movement and evaporation from bare sand dunes, and the uncontrolled flow of groundwater into the sea and saline lakes and marshes. This means that the larger the drainage basin, the more water is lost by evaporation. The ancient farmers were aware of this, and diverted water from small drainage basins to irrigate their terraces, where they grew fruit trees and cereals [3]. These methods, and even the old abandoned structures built for water-harvesting in many places, can be renovated, and the water, which is now wasted, can be utilized for plantations. The diversion of rivers on the margins of the arid and semi-arid countries, which to day flow to the sea, should be considered in the more advanced stages of development. Such are the Oubangui River in the Central African Republic, the rivers of Antalya of Turkey, which flow to the sea, or the Ob River to the Aral Lake basin [4]. Another resource of water is reclaimed sewage from urban centers, which is in increase due to the current world trend of urbanization. Metropolitan centers in arid and semi-arid parts of the world, as well as in the more humid parts of the world, continue to grow. These centers produce large amounts of wastewater, which has to be treated and then disposed of. Utilizing this source for recharge of aquifers, enables its storage, yet in some cases may affect in a negative way the quality of the groundwater. On the other hand the slow flow of the water in the aquifer and its mixing with the natural groundwater, may ameliorate it and enable its use for household utilities, not to speak on irrigation.

References

1. A. S. Issar and R. Nativ, Water beneath deserts: Keys to the past, a resource for the present. *Episodes*, 11(4), 256–262 (1988).
2. A. S. Issar and S. D. Resnick, *Runoff, Infiltration and Subsurface Flow of Water in Arid and Semi-Arid Regions*, Kluwer Academic Publishers, Dordrecht, The Netherlands (1996).

3. M. Evenari, L. Shannan, and N. Tadmor, *The Negev: The Challenge of a Desert*, Harvard University Press, Cambridge, 345 pp (1971).
4. F. Pearce, *When the Rivers Run Dry: Water – The Defining Crisis of the Twenty-First Century*, Beacon Press, Boston, MA (2007).

Chapter 8
Progressive Development by Greening the Deserts, to Mitigate Global Warming and Provide New Land and Income Resources

Arie S. Issar

Traveling back on the dimension of time to the 8th millennium before present (BP) the tourist could have gone for a safari in the present hyper-arid part of the Sahara. He could have sailed over vast lakes of fresh water, watched hippopotami and crocodiles and with the local aborigines hunted the giraffes, gazelles and other savanna type ungulates which roamed the plains [1–3]. One of the French scientists who investigated the paleo-Sahara, late Professor Hugues Faure (Professor of CNRS, France and former chairman of INQUA Carbon Commission, IGCP 404) calculated that during the time when the Sahara flourished, i.e. between 9 K and 7 K BP, about 10 million km^2 of the Sahara were covered by a wooded savanna. In such an environment about 10 kg of carbon were sequestered per square meter by each growing tree. This makes a total of 100 Giga ton (GT) for all the area. As it takes 30 years for a tree to reach the state in which it stops growing which means that if we would have replanted the Sahara, each generation of trees would have used about 3.3 GT of atmospheric carbon each year. This is as much as is needed to sequester all the carbon produced by the fossil fuel which humanity utilizes today (Personal communication by late Prof. Em. Hugues Faure 1998).

Let us suppose, a hypothetical international project to replant the Sahara, the question which has to be answered is where will the water come from to irrigate the trees? At present we know that practically, no rain is falling on this part of the globe. Surprisingly enough water is available under the Sahara to carry out such a project. This water is stored underground in a vast sponge-like ocean, namely sand layers saturated with fresh fossil water. This water was recharged during tens of millennia when the global climate was different. It flows slowly towards the vast saline marshes which border the desert. Only under the Eastern Sahara there exists a volume of about 6×10^{22} m^3. under the Western Sahara 6×10^{13}, Senegal and Mauritania 1.5×10^{12}, Australia 2×10^{13} m^3. Enormous quantities are also found under Saudi Arabia, Kalahari, Gobi, etc. The water is partly artesian and partly shallow enough to be pumped by solar and wind pumps and used for irrigation [4, 5].

A.S. Issar (✉)
Ben Gurion University of the Negev, J. Blaustein Institutes for Desert Research, Sede Boker Campus 84990, Israel
e-mail: issar@bgu.ac.il

A.S. Issar (ed.), *Progressive Development*, Environmental Science and Engineering, DOI 10.1007/978-3-642-10640-8_8,

The order of magnitude of water which is needed for irrigating the 10 million km^2 of trees (considering 1 m^3/m^2/year, although many desert trees need half of that) one would need about 3×10^{14} m^3 for 30 years. Thus enough water is stored under the desert to irrigate vast plantations for many generations.

This being said, the reader should not get the impression that a project is under way to re-savannize the Sahara and thus sequester all the atmospheric carbon dioxide produced by power stations, cars and industry and stop the greenhouse effect altogether. The reason being that on top of the agricultural and technical feasibility, such a project has to prove itself economically feasible which needs further investigation.

Yet, this theoretical exercise may guide us to a new mode of thinking when considering how to mitigate the greenhouse effect. This because besides the hyper-arid parts of the globe which comprises only 7% of the total continental area, about 27% of the world consists of dry lands, including the sub-humid to semi-arid areas (getting seasonal rains below 600 mm/year, and in which the quantity of annual evapo-transpiration exceeds the amount of precipitation).

Over most of these dry lands the density of the population is low, and thus the demand for land is insubstantial, as in these regions, population and agriculture are concentrated in areas where water is obtainable, either from rivers or shallow groundwater. When one looks beyond these restricted, well watered areas, the densities of population and vegetation diminish abruptly and vast stretches of land, in many cases covered by sand dunes or loess deposits, are practically treeless. If only a part of these vast stretches of land could be forested, then large amounts of carbon dioxide could be sequestered.

Many have already pointed out that the creation of a long term carbon sink by tree planting is an important issue for the reduction of carbon dioxide in the atmosphere, and as a matter of fact, is second in its importance, to that of the reduction of emissions.[1] Yet, in the present book an innovative approach to carbon fixation by forestation is suggested, namely of putting emphasis on planting the arid and semi-arid regions of the globe. These are covered today by very sparse vegetation, or are totally bare, but during long periods over the last ten thousand years, were covered by savannas and forests· This is due to the fact that during the periods of proto-history and history our globe went through a few severe natural climate changes, which had their direct impact on the hydrosphere [6]. These changes brought about changes in the bio-phyto habitats and interfered with the history of human societies, especially those living on the margins of the climate belts, which resulted in deforestation and overgrazing [7, 8]. Thus many of the present arid environments which seem to us to be natural, when viewed, from the historical, not to speak of the prehistorical perspective, are just transient states, a result of natural climate changes, and human societies' reaction to it. Such a conclusion puts in a more appropriate

[1] See bibliography in the paper by S. Shea et al. [10] of the Department of Conservation and Land Management, Australia. The Potential for Tree Crops and Vegetation Rehabilitation to Sequester Carbon in Western Australia, Carbon Sequestration Conference 19–21 October 1998, Melbourne. National Association of State Foresters, 1990, Global warming and forestry in the United States, Background paper.

perspective the present Global Change and its impact on the natural environment, as it legitimates affirmative human intervention in order to change what is referred to as the "the natural environment".

The order of magnitude of the quantity of carbon which can be sequestered by replanting these regions is estimated to reach about 30–50% of the total produced by industry. This should be compared with the objective of a reduction of 5.2% emission by the year 2012 (below the level of 1990) which was agreed upon by the politicians and experts, who convened at the Climate Change Summit at Kyoto. Even so, it is still debated whether this modest goal will be achieved, as it means constraints on the rate of growth of the economy of the industrial countries, and opposition groups have already voiced their negative opinion. Without belittling the effort to cut down emissions, the Kyoto conference concentrated its efforts almost entirely on this undertaking which is a non-affirmative action. On the other hand, although the role of plantation as a carbon sink was acknowledged, it did not get the emphasis it should have received. The same omission took place at the Copenhagen Climate Change UN Conference (7-18 December 2009).

As opposed to reducing industrial emissions, planting as a means to reduce CO_2, has merit in terms of long range planning, as it can be described as a "win win" solution. Such a strategy is expedient in any situation when long term planning has to be undertaken, and when there are a variety of possible future scenarios, with which the plans have to cope. In the case of plantations, as a global endeavor to mitigate of the impact of Global Change, society wins as it has good chances to reduce the CO_2 contents of the atmosphere, once the negative future scenarios predicted by the majority of the climatologists and environmentalists are verified. If, however, the predictions are falsified, there is still a gain, this time from the environmental as well as economic aspects, as will be discussed later.

The question to be answered is whether there is enough water in these areas to enable such a project on a global scale? The answer, given in the preceding and the present chapter, is positive. Hydrological and hydro-geological investigations carried out in the arid and semi-arid regions all over the world show that ample water resources can still be found in these regions, and as in the past, human ingenuity will determine whether this water can be utilized to a positive end. Indeed, since the invention of agriculture about ten thousand years ago, farming societies have found ways to make use of the available water resources along the margins of the deserts of the world.

If human society could achieve this when technology was rudimentary, it should be able to make use of the existing water resources, even if they are scarce (due to climate change). In many areas forsaken by past civilizations unable to cope with the worsening environmental and economic conditions, terraced lands lie abandoned; water is now wasted, flowing freely and eroding the soils year by year. Once it is decided to plant these regions, the water available for such projects can come from the different sources mentioned in the preceding chapters.

Of special interest is the subsurface water found in sand dunes, even in areas receiving no more than 100 mm of rain per year. It has been found that in such areas, the moisture content of the soil at the end of the dry season under forested

Plate 8.1 Tamarix forest on sand dunes receiving about 100 mm/year, Negev desert Israel

dunes is higher than that of bare dunes, which emphasizes the positive impact of tree cover on the subsurface water balance of the dunes [9].

Since there is available water, the next question is how the project of planting the deserts should be programmed according to the principles of Progressive Development. These vast areas, which are presently bare dry lands, sustained arboreal vegetation during historical and pre-historical periods. This means that, from the point of view of the ecological system, once the original species are reintroduced, (and this can be done on the basis of the pollen assemblage found buried in the subsurface) re-planting can be a rejuvenation process, rather than the invasion of an alien system. Yet, the knowledge acquired recently that the greener conditions of the past were a function of more humid climate conditions, requires us to put special emphasis on maintaining the hydrological balance, especially when the new plantations will receive their water by irrigation. This balance comprises the quantitative as well as the qualitative aspects of the hydrological balance. In other words, measures have to be taken to manage the resources so as to avoid deterioration of this balance, as for example, by salt concentration in the soil and the subsurface. Such measures include water harvesting methods by terracing, which introduce extra water, otherwise wasted, into the hydrological cycle. This enables the management of the water, soil and tree system on a higher level of efficiency and supply.

To ensure progressive development, projects should be planned and carried out as an integral component of a comprehensive plan of regional development, preceded by hydrological, pedological[2] and ecological research, and followed up by monitoring and re-assessment procedures.

[2] Pedology – soil science.

Another question is that of the economic feasibility of such projects. It would be difficult to dispute the argument that if the plantation of dry lands had been feasible from the economic point of view, then these areas would have been green by now. Yet, at the same time, the recent negative history of the rain forest shows that what pushes many activities of human society are the short term gains, rather than the long term benefits, not to speak of the rather low value assigned to such benefits as biodiversity that cannot be quantified. This being said, it is here argued that plantations in the dry lands can still become profitable enterprises, due to the incorporation of immediate local agricultural benefits, long-term local fringe benefits, and long-term global benefits. However, to encourage the beginning of such enterprises, enlarged subsidies for sequestration of carbon dioxide are required.

For this purpose, special subsidies to encourage such plantations in these regions should be furnished by the atmospheric carbon producing countries. This, in the framework of the steps envisioned to carry out the recommendations of the Kyoto protocol in order to offset carbon emissions.

As discussed in the previous sections, today's impending catastrophe, which endangers the existence of many millions of human beings, is due to global warming, as well as to overcrowding of arable lands by an ever-growing population. Remediation of these two processes, through education and modernization, will take time, and more immediate solutions have to be considered. The basic solution is to target the empty arid regions girdling the globe. These can absorb settlers from overpopulated countries on the one hand, and help mitigate the greenhouse effect on the other. These regions should also be targeted because the land there is of negligible economic value, and last but not least, large quantities of fossil water underlie them and many millions of cubic meters of floodwater are wasted annually, either as flash floods flowing into saline lakes to evaporate, or to the sea. Needless to say that measures to reduce the emission of carbon dioxide in the atmosphere should get first priority, but at the same time, the creation of a long-term carbon sink by planting trees is an important approach, especially because it has many marginal benefits which will be discussed further on, particularly when the trees are being planted in arid and semi-arid regions. Today, these regions are covered by very sparse, if any, vegetation.

It is important to remember that as described early in this chapter, these now empty regions were once green and flourishing. This knowledge is essential to the conceptual model of Progressive Development, as it legitimizes affirmative human intervention in order to change what is referred to by the aficionados of Sustainable Development as the "the natural environment", in which any change is taboo.

The quantity of carbon which can be sequestered annually by replanting these regions is estimated by this author to reach a few tens of percent of the total carbon produced by industry and vehicles. This should be compared with the objective of a 5.2% reduction (of 1990 levels) in emissions by the year 2012, which was agreed upon by the politicians and experts who convened at the Climate Change Summit in Kyoto. Even so, whether this modest goal will be achieved is still under debate, as it means constraints on the rate of growth of industrial countries' economies, and opposition groups have already voiced their negative opinions. No binding Agreement on this issue was reached at the Copenhagen Conference 2009.

A pioneering project in this respect is suggested for the coastal regions of the Negev, Sinai and north Sahara deserts. Thus, although from the narrow commercial and short-term point of view, the production of wood in these regions is not economically feasible, when all benefits, direct and marginal, are considered on a long-term global scale, the planting of woods in these regions has a good chance of being economically viable. The benefits are: payment by the industrial countries for sequestering atmospheric carbon, agricultural products, such as dates, olive oil and other oil-producing trees such as jojoba and jatropha, production of wood for various industries, promoting the biosphere and the vegetation-sphere on the basis of which the tourist industry can develop and thus produce an additional source of income for the people in these regions.[3]

References

1. H. Faure, Lacs Quaternaires du Sahara. *Mitt. International Verein Limnology*, 17, 131–146 (1969).
2. H. Faure, Une importante periode humide du Quaternaire superieur au Sahara – Bul. de l'Institut Fondamental d'Afrique Noire (A). 29/2 (1967).
3. N. Petite-Maire (ed.), *Le Sahara Atlantique l'Holocne. – Alger (AG)*: CRAPE (1979).
4. A. Issar, Fossil water under the Sinai-Negev Peninsula. *Scientific American*, 253(1), 104–110 (1985).
5. A. Issar, and R. Nativ, Water beneath deserts: Keys to the past, a resource for the present. *EPISODES*, 11(4), 256–262 (1988).
6. A. S. Issar, Impacts of climate variations on water management and related socio-economic systems Technical Documents in Hydrology, IHP-UNESCO, Paris, Sc.95/Ws.26, 97 pp. (1995).
7. A. S. Issar, *Water Shall Flow From the Rock. Hydrogeology and Climate in the Lands of the Bible*. Springer-Verlag, Berlin (1990).
8. A. S. Issar, Climate change and the history of the Middle East. *American Scientist*, 83, 350–355 (1995).
9. E. M. Adar, I. Gev, P. Berliner and A. S. Issar, The effect of forestation on a shallow groundwater reservoir in an arid sand dune terrain. *Journal of Arid Land Studies* 58, 259–262 (1995).
10. S. Shea, G. Butcher, P. Bartle Ritson, P. Biggs, Western Australia's Development and Future Prospects for Tree Crop Industries, Australia's Paper and Forestry Forum, 18–19 May 1998, Department of Conservation and land Management p. 18 (1998).

[3]Although it has been recently argued, on the basis of heat-balance models, that "No Climate Benefit Gained by Planting Temperate Forests", as they add shade and reduce heat loss from the surface of the earth (Environment News Service – ENS, 12 December 2006), when it comes to arid and semi-arid regions, it can be argued that although these regions are bare, their cooling effect is negligible, as global temperature increases parallel to the increase in carbon dioxide. Moreover field investigations in Australia and Israel, for example, have measured annual carbon sequestering by trees in arid and semi-arid regions to be between 2 and 4 tons/ha.

Part II
Regional Investigations

Chapter 9
Progressive Development and Groundwater Resources of Israel

Arie S. Issar and Eilon Adar

9.1 Introduction

The history of the development of groundwater resources of Israel is the basis on which the conceptual model of Progressive Development was constructed. This development was dictated by various factors. In the first place stands the socio-economic factor, namely the need to secure a supply of water to answer the demand of a fast growing population due to incoming immigration and natural growth. In the second place come the physical-geographical factors resulting from the fact that Israel is located in a semi-arid zone and is susceptible to abrupt variations in the amount of its annual precipitation, including severe droughts. In addition to this factor is the fact that the average annual amount of precipitation in the southern half of the country is less than 200 mm. The third factor, which was dominant mainly in the past, was the limited funds as well as the educational background of most of the new immigrants which was on the level of many of the countries in the third world at present.

Thus, albeit the term Progressive Development was not phrased at that time the philosophy behind this term evolved simultaneously with the growing needs and resulted in National Water Plan, which will be described in the present chapter.

It goes without saying that the future development of the water resources in this region will have to aim at the supply of adequate quantities to all the population in the region, whether Israeli, Palestinian or Jordanian, yet the issue of future plans depends very much on the way the political issues will be solved. The present study concentrates on the scientific and technical issues of the development of the water resources in the northern and central and humid part of Israel.

A.S. Issar (✉)
Ben Gurion University of the Negev, J. Blaustein Institutes for Desert Research, Sede Boker Campus 84990, Israel
e-mail: issar@bgu.ac.il

A.S. Issar (ed.), *Progressive Development*, Environmental Science and Engineering,
DOI 10.1007/978-3-642-10640-8_9, © Springer-Verlag Berlin Heidelberg 2010

9.2 The History of Water Development of Israel

Pumping of groundwater from wells on a massive scale, started already at the end of the nineteenth century in the Coastal Plain region, to irrigate the orange and citrus orchards. In the beginning the water was raised by an "antillia" type of pump (locally called "noria",) which had a wheel, driven by a donkey or a camel, on which a chain of buckets was rolling. The water was collected in a pond, from where it flowed in ditches, by gravity to the orchard.

The steam-driven motor and piston-pump (introduced by the German Templers, settlers in Palestine) replaced the animal driven pumps. The investment in the excavation and later drilling of a well and installation of a motor and pump was economical only in the case of crops intended for export like the Jaffa orange. Agricultural technology advanced with the arrival of Jewish immigrants from Europe whose academic education included agronomy. From that time irrigated agriculture, mainly from groundwater became the backbone of the economy of Jewish settlements, together with modern agricultural methods, based on scientific research.

During the first 30 years of the twentieth century, cooperative-socialist agricultural settlements (kibbutz and moshav) were established and organized in the framework of cooperative unions on a regional and national scale. These unions backed the settlers politically and financially as well as providing extension services, which involved scientists and agronomists who introduced the most recent agricultural innovations.

In order not to be dependent on the rains for the water supply, wells were drilled and vertical pumps installed, from which the water was transferred by pipes, rather than open ditches. It is no wonder that the first well drilled into limestone was carried out for one of the kibbutzim, namely Gennigar, in the Esdarelon Valley at the foothills of Nazareth. The site was chosen by the authors' teacher, namely Prof. L. Picard (He immigrated from Germany in 1924), who came to the conclusion that the flow in the limestone is under water-table conditions and not restricted to karst solutions channels. The success of this well opened the era of exploitation of groundwater by wells drilled into the limestone and dolomite aquifers. Later exploration wells were drilled into the basalts of the Lower Galilee. After a few failures, Prof. Picard arrived at the conclusion that preference should be given to the drilling along regional fault lines, which caused extensive fracturing of the basaltic rocks. The rate of success enabled the extension of irrigated agriculture to valleys, which up till then relied mainly on rain fed agriculture. The dependence on wells which had to be drilled at selected sites included the need to pipe the water to the various settlements located far from these sites. This encouraged the organization of water supply cooperation on a regional basis.

These initiatives later evolved into regional plans, which necessitated the establishment of the public water company "Mekorot", which in its initial stages planned and executed the projects. The principles which guided the construction of regional projects were extended to inter-regional projects. These involved the supply of water from springs or wells by pipes under pressure, the construction of concrete reservoirs for short term storage, and open reservoirs for seasonal storage.

The integration of regional projects and the conveyance of water from one region to another brought the planners by the early forties to consider a water conveyor from the more humid and relatively water-rich northern part of the country, to the southern drier part. The initiator was the engineer Simcha Blass (later inventor of the drip irrigation method), who in 1944, together with the engineers of Mekorot prepared a plan entitled "Water resources in the Land of Israel: prospects for irrigation and hydro-electric development". The principle of this plan was a water conveyor which would start at the Jordan above the Sea of Galilee and bring water to the southern dry part of the Coastal plain. Experts from the USA like the soil scientist, W.C. Lowdermilk of the soil conservation service of the United States and author of the book "Palestine – Land of the Promise", engineer J.B. Hays and J.S. Cotton, from the Tennessee Valley Authority joined the team of planners. The master plan submitted in 1955, became, after a few changes, the National Water Carrier.

As mentioned the basic idea beyond this engineering enterprise was to transfer water from the more humid northern part of the country, to its dry southern part. Parallel to the water transfer plans, investigations started for developing intensive modern agriculture in this arid region, which was, from this aspect *terra incognita*. For this purpose experimental kibbutz-settlements were established during the forties.

As the area is underlain by impervious chalks the drilling of wells in the vicinity of the experimental settlements failed, yielding only small quantities of saline water. The attempt to catch flood water by dams also failed as this supply was found to be random and thus undependable. At the same time professor Leo Picard located a well field in the north-western part of the Negev, belonging to the Coastal Plain Quaternary sandstone aquifer.

From this well field the first "Negev pipeline", 190 km long and of 6″-diameter pipes supplying 1 million cubic meters per year (MCM) was laid in 1947. This line was converted, during the fifties to a 20″-diameter pipeline, supplying 30 MCM annually. These two projects were followed by the 66″-diameter pipeline of "Yarkon-Negev project", 130 km long constructed during the early fifties soon after the establishment of the State. This pipeline transported 100 MCM annually from the springs of the Yarkon River, emerging from the limestone aquifer of Central Israel in the central part of the foothills region to the Negev over a distance of 108 km.

The last stage of the north-south conveyor was the "National Water Carrier", which started at the Sea of Galilee in the north. The original plan was to draw water from the Jordan River before it entered the lake. The first stages of the groundwork began in 1953. However, because of strong opposition by Syria and a United Nations resolution, Israel was forced to suspend work and modify the initial design. The final plans were approved in 1956, and the National Water Carrier was completed and functioning by 1964. The Carrier is a combination of underground pipelines, open canals, interim reservoirs and tunnels, supplying about 400 MCM annually. Water from the lake, located some 220 m. below sea level, is pumped to an elevation of about 152 m. above sea level. From this height, the water flows by gravitation to the coastal region, whence it is pumped to the Negev. The National Water Carrier functions not only as the main supplier of water, but also as an outlet for surplus water from the north in winter and early spring, as well as a source of recharge to the underground aquifers in the coastal region. Most of the regional water systems are incorporated into the National Water Carrier to form a well-balanced network in which water can be shifted from one line to another according to conditions and needs.

The main artery of the national water enterprise of the State of Israel is the National Water Carrier (Fig. 9.1). Its length, about 130 km, is used to transfer water from the rainy north to the center and the south that are poor in water. Most of the water projects of the country are integrated in the carrier. First plans were begun before the establishment of the state, but, the detailed planning and the execution were enabled only after national independence in 1948 and the project was concluded in 1964.

Transferring water from the North to the Negev was proposed in the year 1939 after the visit of the land preservation expert, Dr. Walter Clay Lowdermilk. In 1944 he published his book *The Land of Israel, The Promised Land* [1] in which he raised the possibilities of agricultural development of the country, and the use of water from the rivers from the north of the country for irrigation in the south. Accordingly a plan was formulated for developing the water of the rivers Jordan, Yarmuk, and Litany. The plan was done by James Benjamin Hayes and was published in the year 1948.

With the establishment of the state, Mr. Simcha Blass set up and directed the Water Section of the Ministry of Agriculture, which oversaw the implementation of the national water project. This was later directed by the TAHAL Company (Water Planning for Israel) which was established in 1952.

The starting point and the main water reservoir of the National Carrier are the Sea of Galilee. From there (in the Sapir site), the water is pumped to the Tzalmon Station and Ilabun Tunnel (the highest point in the National Carrier). The carrier reaches the "Eshkol Reservoir" in the Beit-Netofa Valley, passes through three tunnels, via the Samaria hills and Menashe Heights and from there continues until Rosh Haayin. From Rosh Haayin the carrier water is transferred to the south in pipes of the Yarkon-Negev project. On its way local and Regional water projects are integrated.

The National Carrier was originally planned to supply water only for agriculture needs while the drinking water would be supplied in separated systems, therefore the National Carrier was planned with open channels, open water reservoirs and without the needed treatment to uppercases water in order to supply them for drinking, but after transferring the operation of the carrier from TAHAL to Mekoroth, it was determined that all the water of the Carrier would be in the status of drinking water. The Ministry of Health determined strict standards for the quality of the drinking water. Following this, the Mekoroth Co. operates checking stations for checking the water quality, disinfection systems and water treatment along the "National Water Carrier".

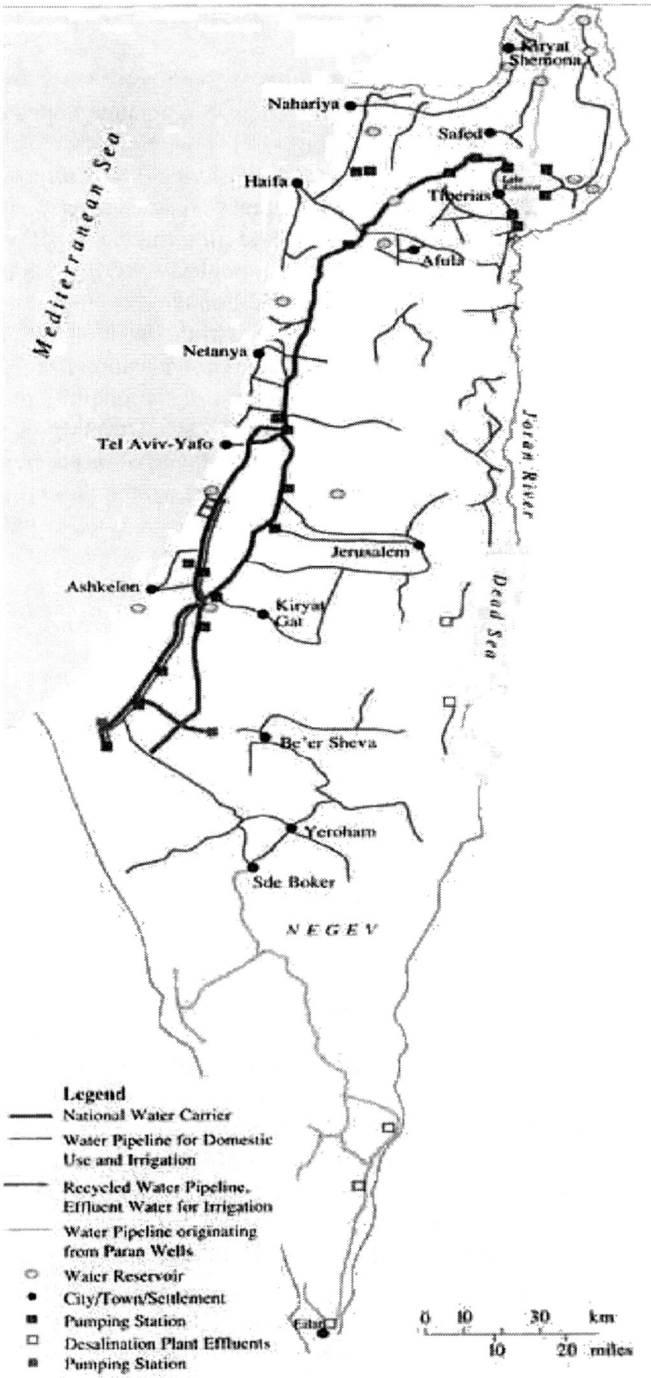

Fig. 9.1 Israel's national water carrier

9.3 Physical-Geographical and Administrative Background

Precipitation in Israel is rather abundant in its northern part, while land resources are rather limited there. On the other hand there is a relative abundance of land in the south where precipitation is low (Fig. 9.2). Thus one of the first endeavors the Government of the State of Israel undertook, immediately after it was established, was to reduce the natural water deficiency in its southern part. This was done, during the fifties, by planning and constructing the National Water Carrier. At the same time the National Water Law was adopted (1959) which declared that "All Water Resources in the country are public domain, are under the ownership of the state and are for the benefit of the people and the development of the country." The law also established the post of a Water Commissioner, and gave him the authority to affect the law and to survey and control the quantity and quality of the water to be used. The law recognized the existing ownership of private people and organizations, before the law was approved, and allowed them to use the quantities they were already using. It also established the regulations according to which the water will be distributed to the various consumers and the tariffs they will pay. The Water Commissioner in consultation with the financial committee of the Knesset decides the prices, which each sector in each part of the country has to pay.

9.4 The Hydrogeology of Israel

9.4.1 The Limestone Aquifers (of Jurassic and Cretaceous Age) Feeding the Jordan River

The base flow of the upper Jordan River, which feeds the Sea of Galilee (Lake Kinneret) comes from the limestone aquifers of Cretaceous and Jurassic age. The recharge area of these aquifers is in Syria and Lebanon. To this lake also flow the winter floods coming from the Galilee highlands and the Golan Heights. From 1980 to 1985, the mean annual contribution of the Jordan River to the lake was 500 million m^3. Surface runoff and wastewater contributed an additional 216 million m^3. The rest came from the direct precipitation and from saline springs at the lake bottom. Of the 812 million m^3 that flowed annually into the lake during this period, 294 million m^3 were lost to evaporation. About 500 million m^3 were pumped for human consumption, and 42 were allowed to flow into the southern part of the Jordan River toward the Dead Sea.

The quantity which can be stored in the Sea of Galilee for pumping, is limited to about 600 million m^3. The low-level constraint is due to the existence of saline springs at the bottom of the lake, the flow of which may increase if the level of the lakes drops too much. On the other hand maintaining a high level in wet years, by keeping closed the gates at the southern outlet of the lake, may cause the flooding of the city of Tiberias and other settlements along the banks of the lake.

Fig. 9.2 Precipitation map Israel

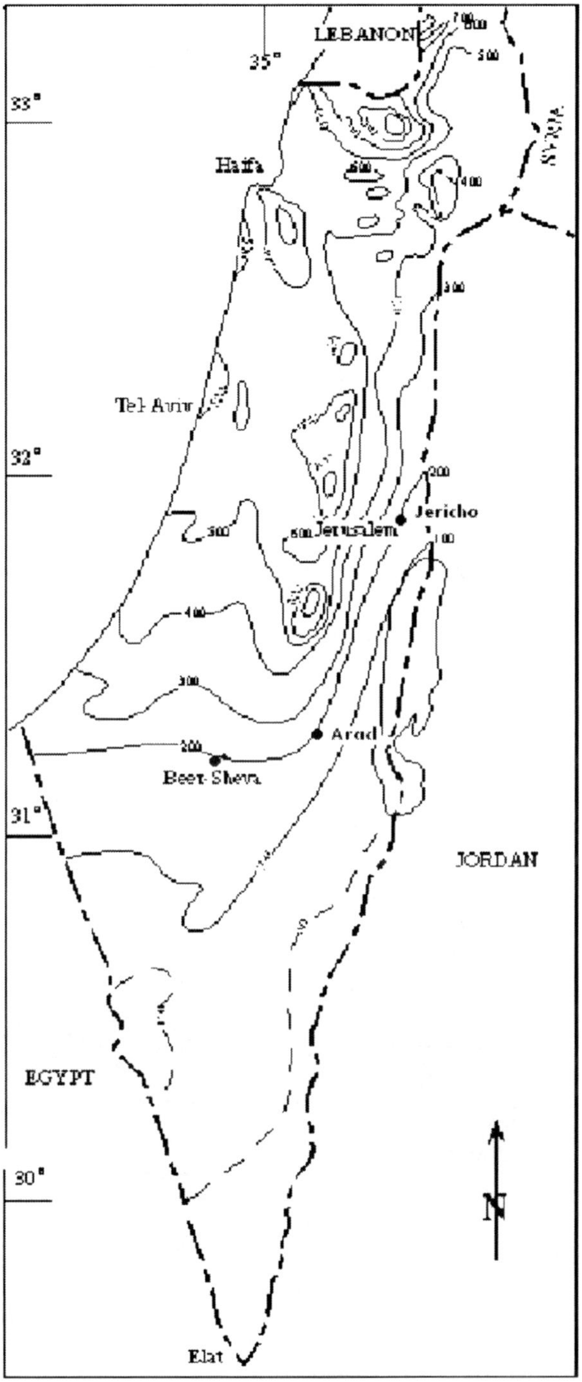

9.4.2 The Cretaceous Limestone Aquifer of Central Israel

The other limestone aquifer is built of the Judea Group (Turonian-Cenomanian Middle Cretaceous age), which builds the Galilee Mountains. The water infiltrating flows vertically until it reaches the saturated part of the permeable limestone, from there it flows in a sub-horizontal direction, as groundwater, in all directions. In the Galilee, the Judean Group aquifer is subdivided into upper and lower parts by a thick sequence of impermeable chalks. The lower aquifer is mostly confined. A regional groundwater-divide which divides the Galilee into two main groundwater provinces is located more or less along the topographical backbone of the region. The discharge to the east is mainly by springs, the average annual flow of which is about 110 MCM. Another 30–50 MCM are pumped annually. Part of the groundwater flow emerges as saline springs, part of it at the bottom of the Sea of Galilee (on the mechanism of salination (see: Issar, 1993)). The saline springs along the shore are diverted to flow directly to the Jordan River below the lake. The volume is about 20 MCM. This water is planned to be desalinized and delivered to the Jordanians (Hydrological Service Report, 1998). The flow to the west, discharged mainly by wells and also by springs, is about 70 MCM.

In the central part of the country the subdivision is of a local nature, and this aquifer can be regarded as one hydrological system. All the western and south-western flank of the mountainous part forms one regional aquifer (termed Yarkon-Taninim from the main springs in its outlet). In the foothill regions there is practically no recharge as the limestone is covered by impermeable layers, chalks and marls (Upper Cretaceous, Paleogene and Neogene age). In this region the water table is sub-artesian (confined), and rises to a certain level, when the borehole strikes the top of the aquifer. The average annual recharge of the Yarkon Taninim aquifer is about 360 million m^3. Increased water pumping from the aquifer reduced the western natural discharge to about 20 MCM of brackish water (Fig. 9.3).

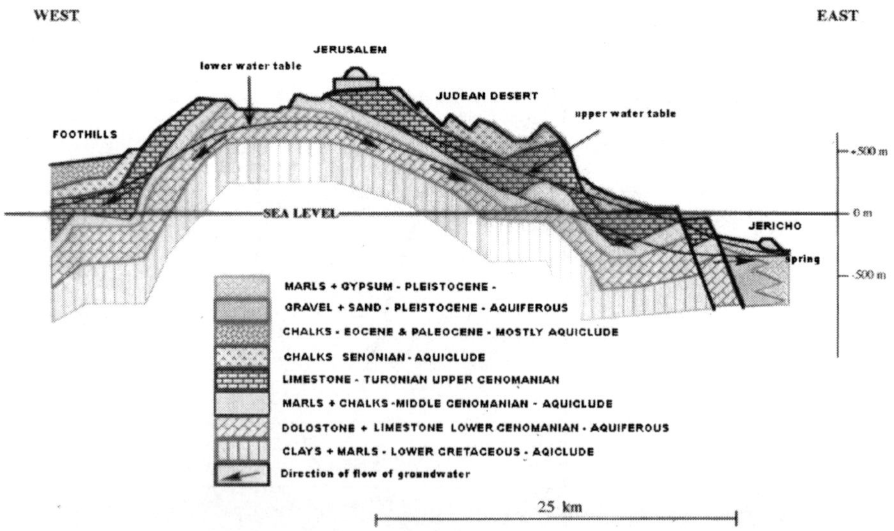

Fig. 9.3 Geo-hydrological cross section over Central Israel

In the eastern part of the mountainous region the groundwater in the limestone aquifer flows towards the Rift Valley. The quantity that flows to the northeast and east is about 250 MCM (Hydrological Service Report, 1998). Along the eastern coast of the Dead Sea saline springs are found which get their salts from contact with the interface of the Dead Sea water.

9.4.3 The Sandstone Coastal Plain Aquifer of Quaternary Age

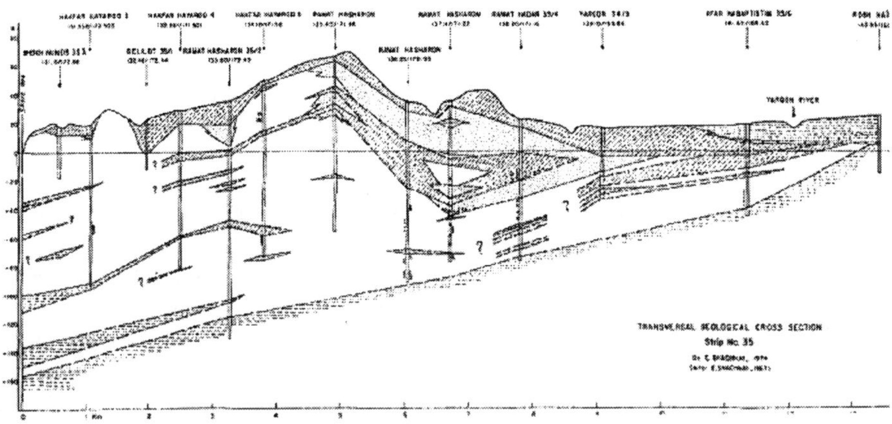

Fig. 9.4 Cross section east west Coastal Plain

The Coastal Plain aquifer is built of permeable sandstone rocks with inter-layering of semi-permeable loams to impermeable clays. In the western part of region the impermeable layers divide the aquifer into sub-aquifers; the upper one is phreatic, while the lower sub-aquifers are confined. Due to this subdivision, recharge and pollution in the west affects only the upper-most sub-aquifer. In this part of the Coastal Plain, all along the coast, there are areas in which; due to over-pumping and decline of the groundwater table, there is penetration of the sea interface. This, however, is differentiated according to the position of the hydraulic head in each sub aquifer. The separating layers disappear a few kilometers away from the shoreline towards the east, and the sub aquifers merge together to form one phreatic system. Thus, in the central and eastern part of the Coastal Plain the aquifer is one unit, being subdivided only by semi-permeable loams, which retard but do not confine vertical flow.

The annual natural recharge during 1996/1997, of this aquifer, was calculated by the Hydrological Service (1998) to reach 242 MCM, while the return flow from irrigation reached 57 MCM. About 43 MCM came from the east, while about 110 MCM was recharged artificially. In total about 451 MCM were recharged, while 407 MCM were pumped out. About 38 MCM flowed to the sea. Generally, the groundwater flow in the Coastal Plain aquifer is from east to west except in areas where there exist cones of depression due to massive lowering of the water table because of excess pumping. In these areas the flow is directed towards the groundwater table depressions.

The salinity of the water in this aquifer is 50% higher today than it was during the thirties. This is due mainly to the fact that the water in this region was used mainly for irrigation. Nitrates have increased from an average of 10 mg/l in the 1930s to more than 50 mg/l at present. In the southern half of the Coastal Plain its sandstone layers are in contact with semi-permeable chalk layers of Neogene and Paleogene age. In these areas there is an infiltration of saline water, this causes the water in the southeastern part of this aquifer to become brackish (400–800 mg/l Cl) and some area to contain high levels of natural nitrates (30–70 mg/l).

9.4.4 The Limestone and Chalk Aquifers of Eocene Age

This aquifer, although of minor importance from the quantitative point of view, is of significance as it causes salinity in the eastern parts of the Coastal plain aquifer. In the Galilee as well as in Mount Gilboa it is found in a reef facies (Bar Kochba Limestone). The total annual quantity discharged from this aquifer either by springs or pumping is in the order of magnitude of 100 MCM. The salinity is low.

In the northwestern Negev, the aquifer is built of partly dissolved fractured chalk. The salinity may reach 600–10,000 TDS mg/l the source of which is ancient residual seawater trapped in the chalk pores. Pumping tests performed on the wells in the aquifer indicate aquifer transmissivities from 0.01 up to 100 m^2/day.

9.5 Advanced Methods of Irrigation

In Israel, the agriculture sector is the major consumer of water. Thus, to curtail the total water consumption, the amount of water allocated to agriculture has been subjected to a number of restrictions, especially since the early 1990s. From a total consumption of 2,008 MCM in 1997, 1,264 MCM (63%) was used for agriculture, as compared to the situation in 1985 when water consumption for agriculture was 1,389 MCM out of a total of 1,920 MCM (72%). There is no doubt that efficient use of water for irrigation is a paramount priority.

One of the most important agro-technological innovations is probably the invention in Israel of drip irrigation. The principle was conceived in the UK, Simcha Blass developed it to the level of a field irrigation system while his son further developed the dripper. The members of Kibutz Hatzerim, near Beer Sheva, which faced the problem of having brackish water and saline soils, promoted it to become an international industry.

Drip irrigation enables uniform irrigation even on slopes, while the water and fertilizers are economized, being delivered directly to the root system rather than to the total area of the field. This method makes it possible to utilize brackish water, as it causes the salts to be continuously washed away from the root system. As the supply of water by the drippers is limited to the close vicinity of the root system of the plant, the spray of salt on the leaves is avoided and at the same time the efficiency of the irrigation, i.e. the ratio between the amount of water taken up by the plant and the total amount of water applied, is optimal.

Even with all the innovations, there is still a shortage of water in Israel in recent years, due to global warming, which has caused a sequence of drought years as forecasted by the authors. At the same time in the last ten years there was an increase of the population by about 31% (from 4.8 million in 1990 to 6.3 million in 2000), due to the collapse of the USSR and the escape/immigration of Jews. This brought about an increase in the demand for water for irrigation as well as for domestic use.

9.6 The Reuse of Sewage Water

Increasing quantities of sewage water are being generated, endangering groundwater and other sources of fresh water. The pressing need to find alternate sources of water, together with the critical condition of the environment, led the Water Commission to set up the Shafdan plant, a large-scale project for processing sewage to produce highly purified water. This procedure results in two major benefits: the aquifer serves as an underground reservoir for the recharged water – preventing losses by evaporation – and water is pumped off when needed, mainly in summer; percolation of the water through soil layers provides an additional cleaning phase.

About 110 MCM of this purified water is transported annually via a separate pipeline called the "Third Negev Pipeline" to the western Negev for use in irrigation. Thanks to the high degree of purification of the treated water, it can be used for all crops without risk to health.

Additional sewage water purification plants are already under construction or are in the planning stages. It is expected that most of the water allocated for agriculture will eventually consist of purified effluents, so that quality fresh water can eventually be shifted from agricultural to domestic uses.

Smaller-scale plants all over the country provide treated sewage water for irrigation of fields located a short distance from the source of the effluent. In many cases treatment is minimal, and use of the treated water is restricted to crops such as cotton in the summer. However, small projects of this type are reported to be highly cost-effective.

Altogether in the State of Israel, 450 MCM/year (million m^3) of treated sewage drainage are produced. About 290 MCM/year of treated sewage waters per year, (about 64%) are returned for use in agriculture with the help of sewage purification institutes.

Most of the sewerage flows into the drainage and conductance systems (96%), but 4% are absorbed in the soil in cesspits.

Altogether 63% of the drainage (283 MCM/year) is treated at a proper level, while 37% (160 MCM/year) is not treated at all or only at a very low level.

9.7 Use of Brackish, Saline and Seawater

Brackish water, namely of salinities above that enabling its use for drinking purposes, is already being used for irrigation of trees (palms, olives) and crops (cotton, tomato, melon) tolerant to a certain quantity of salts in the water (2,400–6,000 ppm TDS). As will be discussed later, in order to economize on water on one hand, and to avoid accumulation of salts in the root zone and even leach away the salts below it, on the other hand, the method of drip irrigation must be applied.

When the salinity is beyond the tolerance of the vegetation, the water has to be desalinized. In many parts of the country there exist other resources of saline water, which cannot be used directly, but have to undergo desalination. Reverse osmosis, in which the saline water is pushed through special membranes allowing only fresh water to pass through, was found to be the most efficient and economical.

Desalination of saline water is preferred to desalination of seawater, since the energy required to produce drinkable water from saline water is 0.8–1.0 kWh/m^3, and 73% of the water input is recovered, while the energy required for desalination of seawater is about 3.85 kWh per cubic meter, and only 50% of the water input is recovered. As the average annual replenishment is about 1.5 BCM (billion m^3), while the annual total consumption is about 2 BCM, the annual deficit is 500 MCM. Moreover there is an accumulating deficit of more than 2 BCM. In addition to the quantitative deficit there is an ongoing process of deterioration of the quality of the water. In order to deal with the deficit and deterioration, a national endeavor was undertaken to develop second grade water sources, namely reclaiming wastewater as well as intercepting runoff mainly for artificial recharge. Since 2005 desalination of brackish and seawater has begun. During 2008 about 150 MCM of seawater were desalinized. The desalination plants under construction will enable the desalination of 280 MCM in 2010, while by 2013 the quantity of desalinized water to be reached is 500 MCM.

References

1. W. Lowdermilk, *Palestine – Land of Promise*, Victor Gollancz, London (1944).
2. A.S. Issar, Recharge and salinization processes in the carbonate aquifer in Israel. *Environmental Geology*, 21:152–159 (1993).

Chapter 10
The Negev Desert of Israel – A Conceptual Plan of a Progressive Development Project for an Arid Region

Arie S. Issar and Eilon Adar

10.1 Geography, Climate and Geology

The Negev Desert is situated on the intersection lines between climatic zones, geological provinces, ecological systems, and human societies. Due to its special situation one can find in it most types of desert environments, on most scales of aridity. These environments comprise rolling sand dunes, mountainous rocky terrains, loess plains, savannas, alluvial fans and salt marshes. While the Mediterranean climate affects its northern part, its eastern part is affected by the down faulted rift valley ("graben") topographic features of the Arava Valley, part of the Syrian-African Rift system. This major rupture in the earth's crust, a geological embryonic phase of a new ocean, contains the Dead Sea in its deepest section, and the Red Sea in its most southern part. Since it was formed, just a few millions years ago, it has caused the deep gorges which flow into it, to cut down and expose a geological section from the most ancient to the youngest layers. In the rocks one can find the most ancient fossils, while in the river terraces the most ancient artifacts. The various types of rocks also produce all land forms of erosion.

Lying between the lands which were the cradles of human culture, Egypt to the southeast, and Mesopotamia to the northwest, the Negev Desert has formed, since prehistory, a theater of the conflict between nomadic and agricultural societies, between the city dweller and its pillager.

At present, extensive parts of the Negev Desert like other regions along the margins of the Sahara-Arabian Desert are undergoing desertification.[1] The reasons for this process are natural as well as anthropogenic, each ameliorating and even accelerating the other.

A.S. Issar (✉)
Ben Gurion University of the Negev, J. Blaustein Institutes for Desert Research, Sede Boker Campus 84990, Israel
e-mail: issar@bgu.ac.il

[1] This term, which comprises also the socio-economic aspects of the process, will still be used in the following chapter in spite of the fact that UNEP conference in 1990 in Nairobi agreed that land degradation is a more useful term.

A.S. Issar (ed.), *Progressive Development*, Environmental Science and Engineering, DOI 10.1007/978-3-642-10640-8_10,

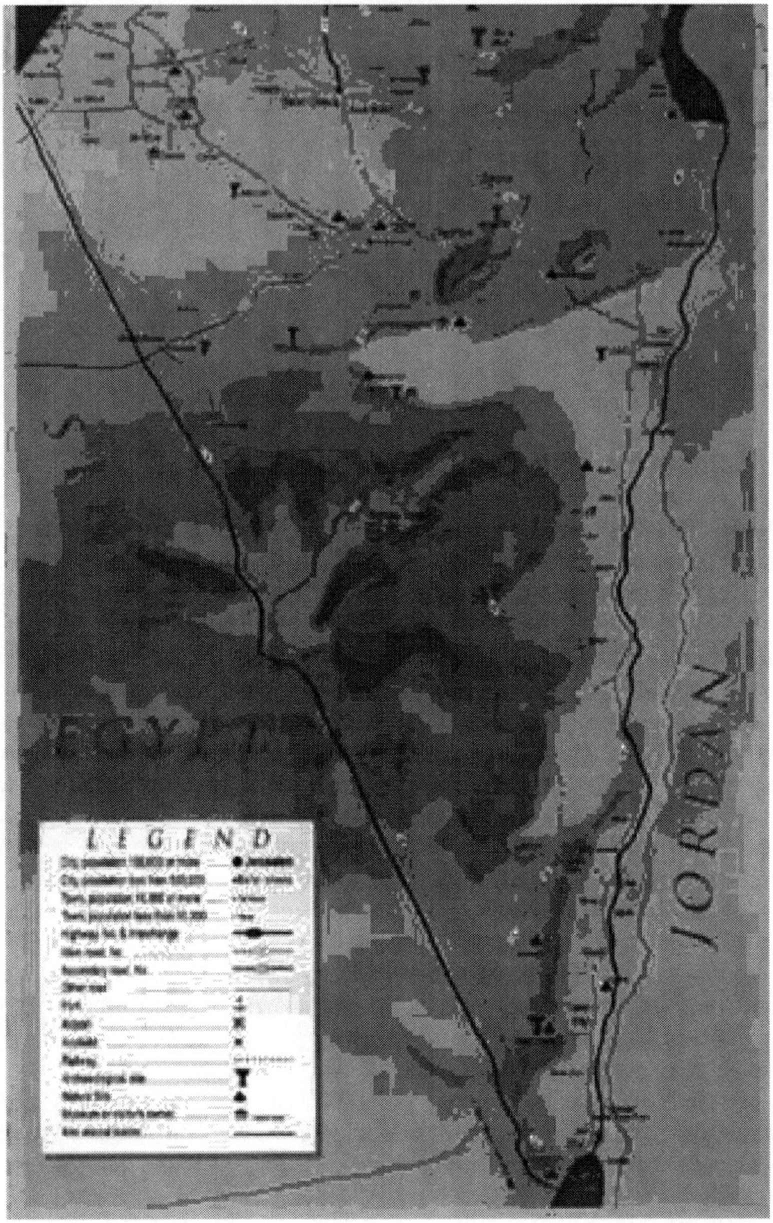

Fig. 10.1 A morphological map of the Negev

The natural reasons are connected with droughts, which are due to the present climate change. These are expressed by the dynamical processes in nature, like the advance of sand dunes on one hand and the erosion of soils on the other. The anthropogenic processes are a function of the droughts as well as the high rate of increase of the population of the Bedouin semi-nomadic population. Combined, these processes drive the process of overgrazing and cutting the natural vegetation for fuel wood. These latter processes exceed the natural renewal capacity of the vegetation and soil, which brings about its denudation.

The expansion of the deserts, including that of the Negev, works against the global need of enlarging the supply of food to the world population, especially the populations living along the boundaries of the deserts. This need can be partially answered by turning areas which contain unused water, either groundwater or wasted floods into productive land. This recommendation is based both on the results of many investigations which show that such water resources exist, and on the observations that many of the areas considered today as deserts, were fertile countries during historical times, for example: the Maghreb lands of Africa were the "bread basket" of Rome. Many ruins of once beautiful cities surrounded by thousands of hectares of abandoned terraced river beds are known from the countries of the Levant [1, 2]. Whether the reason for the desertion be climate or man, or both climate and man, the fact is that once these lands supported an agricultural population, and today they lie desolate, benefiting no one, and a nascent area for the slow but sure progress of the desert. It can thus be stated generally that many deserts, including the Negev Desert, are potential areas of food production and await development.

10.2 The Past as a Key to the Future

Due to the arid climate in the Negev Desert the remnants of the endeavors of its ancient inhabitants to cope with the natural conditions for securing a living were preserved. As this region was, since the arrival of the hominids, on the margins of the desert, these remnants tell the story of the climate changes which caused this region to become more humid or more dry, yet at the same time one finds evidence of human creativity and inventiveness to find ways to ameliorate the benefits of nature, bestowed on this region in paucity.

At some stage of the aridization of the margins of the deserts, part of the inhabitants adapted to the fluctuations of supply and adopted the nomadic way of life, which enabled them to find enough forage for their livestock, on which they became dependent. In the nomads' societies the number of heads in their herds decided their status and chances for survival in bad years. When the number of livestock was small, water could be stored in cisterns and reservoirs, but when the number increased, the use of wells dug to reach groundwater had to be introduced. The gradual development of the technique of excavating deeper wells for tapping groundwater enabled man to penetrate deeper into the desert where his livestock could graze in the pasture.

Yet when climate was optimal enough to enable farming like during the Chalcolithic Period (4500–3300 BCE) the inhabitants developed an irrigation method known today as water harvesting. This method included building low dikes of soil and stones that crossed the small creeks and held back the floodwater which could then slowly infiltrate the ground. The floods also brought with them rich soil from the surrounding hills that fertilized and enlarged the fields or the orchards. Because the area was barren in summer, they dug shallow cone shaped wells, to supply drinking water for the inhabitants and their livestock [3].

Cold and humid conditions occurred also during the Early Bronze Period (3300–2200 BCE) [4, 5]. A water supply system originating from a central well, and a system for enriching groundwater was found in the city of Arad, in the arid part of Israel, on the northeastern border of the Negev. The city was built in such a manner that all its streets led towards a pool and a well. When heavy rain fell, the built up area became a collecting device for the water that flowed along the streets to a shaft dug at the lowest point in the city. The water in the shaft enriched the shallow perched groundwater table and later, during the dry season, water was pumped from the shaft-well to supply the city. Due to the limited recharge area, the reserve of this perched groundwater table was quite limited, and during the warm dry periods which followed it dried up and the city was deserted.

The studies of U. Avner [6] show that, as in the Chalcolithic Period, the Early Bronze Age settlements flourished in the central part of the southern Negev desert, such as the valley of Uvda. These agricultural settlements used various types of dams, harvested water for irrigation, and excavated wells into the perched groundwater table.

The following period known as the Intermediate Bronze period (2350–2200 BCE), was very dry and spelled disaster for all of the Middle East [4, 5]. The farming societies deserted their farms and were replaced by the semi-nomads.

The extreme aridity of the Intermediate Bronze Period was followed by the temperate climate characteristic of the Middle Bronze Period (2200–1200 BCE). Yet, the improvement in the climate was not enough to revive the agricultural settlement of the Negev.

The Iron Age (1200–580 BCE) was also rather cold and humid. The city of Beer-Sheva on the northern border of the Negev flourished as an administration center. The city was circled with a thick wall built of bricks. Inside the walls a deep shaft with spiraled stairs which led to an undergroundwater reservoir which was fed by a diversion canal from the Beersheba Stream. Outside the walls near the main gate a well, 70 m deep, was excavated.

South of Beer-Sheva in the desert itself many water reservoirs were dug during this period. These uncovered reservoirs are located on an impermeable marl layer and get their water from a sloping canal circling the hill. That the inhabitants also practiced agriculture can be learnt from the existence of terraces, which were built in the riverbeds.

The Hellenistic – Roman – Byzantine Periods (300 BCE to 650 AD) were also relatively cold and rainy periods. During this period Nabataeans began to settle in the desert in the southern part of the Fertile Crescent. The agricultural system of these settlements was based on water harvesting which is the collecting of water from small drainage basins on the mountain slopes to irrigate the agricultural terraces built in the surrounding valley. The team that investigated this method [2] did not take climate change into account and assumed that the average yearly rainfall was the same as it is today. In fact, the yearly precipitation at the time was 30–50% more which explains the flourishing agricultural terraces all over the area bordered to the south by the present 100 mm line. Numerous remnants from the ancient agricultural settlements of these periods have been found.

The ruins of six deserted ancient cities (Avdat (Abde), Halutza (Halassa), Mamshit (Kurnub), Nizanna (Auja el hafir), Shivta (Subeita), Rehovot (Ruheiba)) are also found in the Negev. The richness of these cities was, according to various historians and archaeologists, related to the fact that they were situated on the main routes of commerce, especially of frankincense brought from southern Arabia and shipped through the Mediterranean ports to Rome. Yet the widespread agricultural farms containing wine and oil producing installations, which surround these cities, show that agriculture and its commercial products were part of the richness of these cities.

Plate 10.1 Wine press Avdat (Photo. A. Issar)

The farmers, supplying the agricultural products lived in stone built houses in the vicinity of the riverbed which they terraced. Each house had one cistern close to the house and a few more spread over the property. In these cisterns floodwater from the hill slopes was stored. Prior to its flow into the cistern the mud was deposited in a retention tank.

The farmed land consisted of terraced riverbeds. The terraces were built of stone walls (a few tens of meters long in each river bed and a few meters high) forming low dikes at right angles to the direction of flow. These terraces slowed the flow of the floods and caused them to inundate the entire river bed, causing the deposition of mud load. Thus irrigation and rejuvenating the soil were carried out simultaneously. The runoff from the hill slopes was diverted to the terraced riverbeds by a system of channels. The total area of the runoff harvesting area to the irrigated area was about 20:1 [2]. The agriculture system surrounding these settlements was called "Water Harvesting" method. This method was based on catching the surface water flowing over the bare, rocky slopes of the hills. It also included diverting the floods from main riverbeds, rather than damming the entire river flow. The people who planned and built these systems were aware that flash floods in the desert are difficult to control because of their unpredictable volume and sudden harshness. Experience has taught them that desert flash floods are especially severe because there is no soil and tree cover on the hillsides to mitigate their impact. In addition, intense cyclonic rainstorms bring a tremendous downpour to a limited area within a rather short time and the resulting floods are usually extremely intense and destructive. Thus the method was based on collecting the waters from small drainage basins that were extended on the mountain slopes that surrounded a valley where agricultural terraces were built across its channel. The expanse of the basin from where the waters were collected into canals that were excavated in a moderate slope on the downgrades of the mountains extended from tens to hundreds of hectares. The canals led the waters to the creek or riverbed, which was terraced, and collected the mud brought by the surface flow to become agricultural soil. As mentioned the sub-parceling of the slopes and dividing it according to the area of terraces to be irrigated, upgraded the efficiency of the irrigation, because as long as the land parcels were smaller, so was the ratio higher of the flow to that of the rain falling on the parcel. The ratio between the area of the collecting basin on the inclinations and that of the cultivated terraces is approximately 20:1, meaning, that every hectares of the farms land needed about 20 ha of slopes in order to collect enough water for irrigation.

The Evenari team that investigated this method did not take into account climate changes, and assumed that the yearly rate of precipitations averaged the same as the present, namely around 100 mm. Therefore for an average of 15% flow, each hectare is producing an average of 150 m^3, therefore 20 ha will produce about 3,000 m^3 of water, to irrigate 1 ha, which is equivalent to about 300 mm of rain. If we add to this the natural 100 mm of rain then the total is about 400 mm/year. Discussing this issue with the local farmers, Issar was informed that this quantity is sufficient to support the plantations without causing them to degenerate, but it does not ensure good crops. Once climate change is taken into account and that the quantity of rain was 30% (and during good years even –50%) more than the current quantity then the average was enough to ensure an optimal crop. This, jointly with the fact that there were certainly also fewer drought years, might explain the fact that the earlier inhabitants of this arid area did not exclude even the smallest river basins from terrace building to cultivate them.

Another feature peculiar to the Negev is *Tuleilat el Anab* (grape mounds in Arabic) which are spread over the hill slopes. The assumption by most researchers was that they were intended to enhance the flow of run-off water into agricultural terraces. Yet, this assumption did not take into consideration the fact of climate change. The explanation of the present authors is that these were used for growing

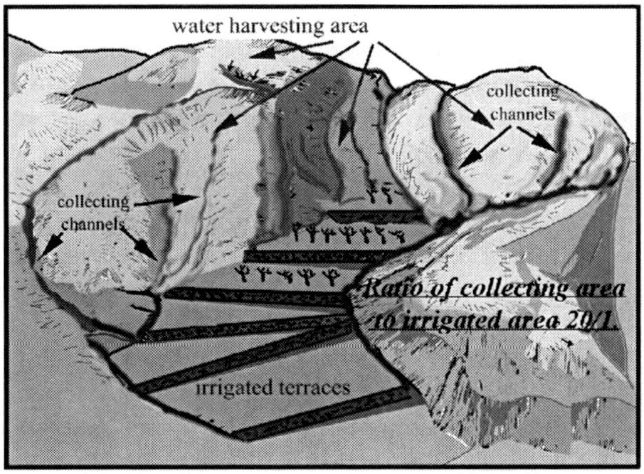

Fig. 10.2 A water harvesting system in the Negev

vines, which covered the mounds got some water from the runoff on the slope and some from the dew which dripped from the stones of the mound directly to the grape vines. One finds a very similar way of growing vines on the island of Santorin.

The flourishing of the agricultural settlements based on water harvesting continued from the Nabatean period (second century BC–second century AD) and continued through most of the Byzantine period, to the Umayyad period and continued until the middle of the eighth century AD. The extreme climate crises evidenced by the numerous proxy data available today (stalagmites, Dead Sea and lake Van levels, soil profiles) effected the desertion of all the cities and farms in the southern part of the Levant, including the Negev. It can be thus concluded that during historical times, settling in the desert and desertion of settlements was a function of severe climate changes, triggering not only man's ingenious invention, but also failure and misfortune.

During the last 60 years, the Negev has undergone a new phase of settlement not due this time to climate change, but to national investment in order to alleviate the mounting population density in the Coastal Plain.

10.3 Groundwater the Basis of Development of the Negev

It is quite obvious that the achievements of the ancient inhabitants in the technology of water harvesting could last as long as the multi-annual average of precipitation did not drop below a certain level. Once this happened, either due to a general decrease in the average or due to a re-occurring series of droughts the sophisticated agricultural system collapsed. The drying up of the whole southwestern part of the Asian continent as well as Mediterranean part of Africa brought in its wake the turning of this flourishing region into a desolate land. On the other hand the regions, which got their water from springs emerging and fed from regional aquifers, like Mesopotamia, survived the climatic crisis. Thus, although in general terms it can be stated that on the whole there exist ample resources of unutilized runoff water in the Negev Desert, this resource is unreliable as the basis for a modern economy, including modern agriculture. The alternative resource is groundwater.

Numerous investigations have proved that despite being a desert region groundwater resources still exist under the Negev Desert, which allow the planning and development of this region, and which are able to supply the needs of an extensive population enjoying a high standard of living.

Four main aquifers containing fresh to brackish water underlie the Negev Desert. The most prominent one is the limestone – dolomite aquifer belonging to the Judean Group of Cenomanian-Turonian age. The second in importance is the Nubian Sandstone aquifer of Lower Cretaceous age. The third is the alluvial fill of the Rift Valley. The fourth is the limestone aquifer of Eocene Age.

In general it can be said that the two first aquifers are part of a regional aquifer extending below the Sinai Desert. The general direction of flow is from the outcrops bordering the igneous block of southern Sinai towards the Suez Gulf in the west and the Rift Valley and Dead Sea in the east.

10.3.1 The Limestone – Dolomite Aquifer of the Judean Group of Cenomanian-Turonian Age

The average thickness of the limestone-dolomite aquifer is about 500 m. Its salinities range from 500 to 2,000 mg/l Cl. The water is mainly fossil; its age range is around 10 K years.

In the anticlinal regions, where this aquifer outcrops, recharge takes place and local groundwater mounds are formed in which water is younger and less saline. However, the quantities of contemporary recharge are negligible in comparison with the general quantity of fossil water stored in the aquifers. The directions of flow are decided by the main anticlinal structures and fault lines. The most prominent elevated structure is the Ramon anticline crossing the central part of the Negev laterally. In the northern part of the Negev two smaller anticlines extend. In the anticlinal zones the rocks of these aquifers were eroded and thus these anticlines form subsurface dams which divert the flow into the synclines extending parallel to them. A regional fault line, which influences the direction of flow, is the Paran fault.

This aquifer is divided into two provinces, one to the north and the other to the south of the Ramon anticline.

In the northern province the flow is in two main directions i.e. from the border of Sinai, (the Nizana region) towards the north. South of Beer-Sheva it interconnects with the main aquifer of the central part of Israel, i.e. the Yarkon Taninim aquifer. The other direction of flow is towards the Dead Sea in the east, and south towards the Makhtesh Gadol and Makhtesh Katan.

In the southern province the flow direction is from the border of Sinai to the Arava Rift Valley. In this province the flow is divided by the Paran fault. On both sides the main flow is towards the northeast into the Arava Rift valley.

10.3.2 The Nubian Sandstone Aquifer of Lower Cretaceous Age

The average thickness of this aquifer is about 250 m. The salinity varies between 800 and 2,000 mg/l Cl. The water is fossil; its age range is around 20 K years. In the northern part of the Negev there is not enough data to find out the direction of flow. It can be assumed that if there exists a flow, it is mainly to the east, i.e. to the Dead Sea. In the southern part of the Negev the direction of flow is parallel to that of the Judea Limestone, namely from the highlands of Sinai, where these rocks outcrop, in a northeast direction to the Rift Valley.

10.3.3 The Alluvial Aquifer of the Rift Valley

Its thickness varies from a few tens to a few hundred meters. Recharge is contemporary, salinities are mostly low. The total annual recharge is in the order of magnitude of 30 MCM, while the quantity pumped is about 60 MCM, of which half is mining of fossil water.

10.4 Modelling

In the frame work of a regional research program,[2] the general conceptual model was transformed into a computerized flow model.[3] This involved:

1. Location of borders and decision as to boundary conditions.
2. Division of the region into layers according to geology.
3. Definition of inflow, pumping and outflow areas.
4. Construction of a Cartesian three-dimensional grid system, in which each layer is represented by a two-dimensional grid.
5. Definition of the hydraulic parameters for each layer (permeability, storativity).

For the simulation of the model a GMS program package was used, running MODFLOW for groundwater flow simulation.

The model was calibrated for steady state as well as for dynamic conditions. The calibration was successful for all parts of the Negev except for the region extending for about 40 km south to the Ramon anticline, most probably due to the scant hydrogeological information in this region. After calibration, the flow model was run to simulate the response of the limestone and sandstone fossil aquifers to long-term

[2]Zuckerberg Institute for Desert Research, 2004, "The Pratt Research Project, Treatment and usage of brackish water. The feasibility of desalination of local brackish/salty groundwater and wastewater treatment versus importation of desalinated seawater Beer-Sheva basin and the Negev Highlands".

[3]Tahal Consulting Engineers Ltd and A. Issar.

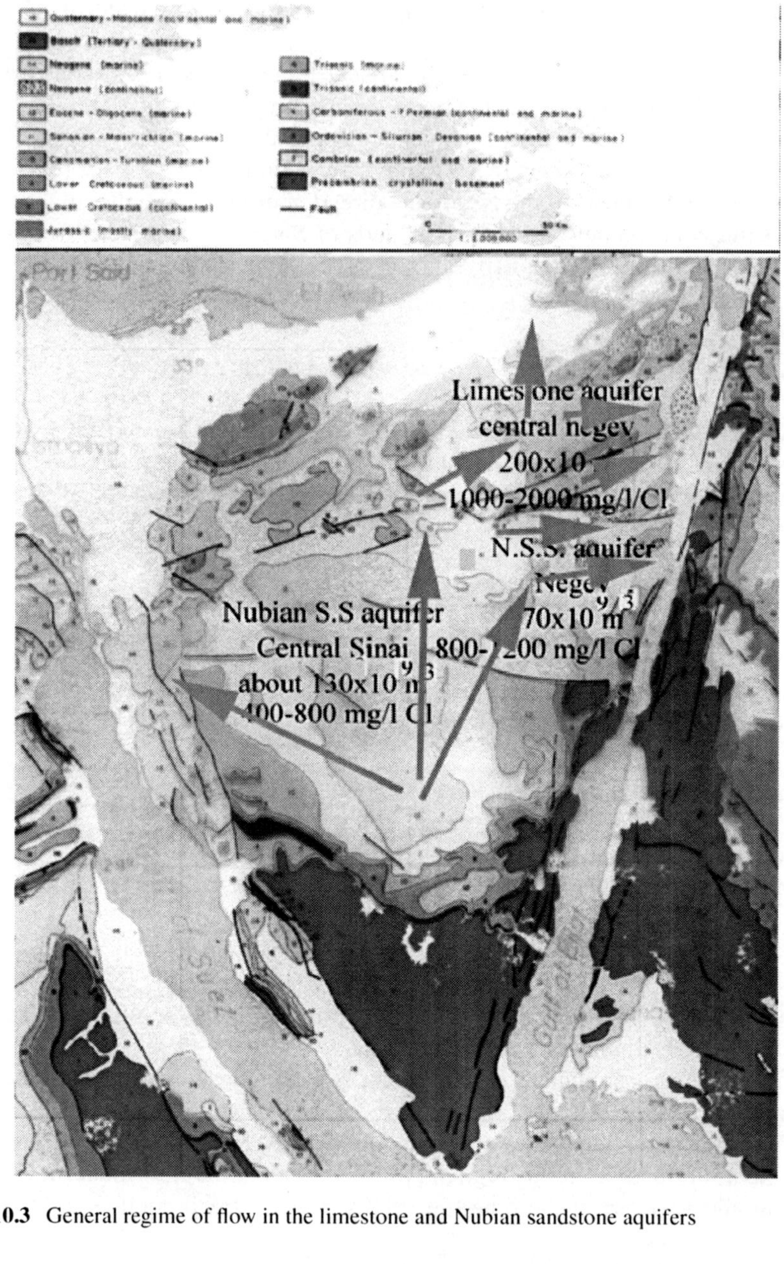

Fig. 10.3 General regime of flow in the limestone and Nubian sandstone aquifers

excessive utilization. The simulation of pumping of about 100 million m^3/year from each aquifer resulted in a total draw down of the order of 50–60 m, i.e. an average of 1.6–2 m/year in the Judea Limestone aquifer and 50–100 m, i.e. 1.6–3.3 m in the N.S. aquifer. In total about 200 MCM/year can be pumped from these aquifers for the next 100 years. After this period the water table will be lowered by 200–400 m, yet in most areas there will be still ample quantities of water left in these aquifers. With regard of the cost of the water which will have to be pumped from greater depths, it should be taken into account that a century from today the advance of technology of pumping, irrigation and agricultural methods will enable the future consumers to pay for the water.

10.5 Regional Plan of Development

The Negev Desert occupies about 60% of the total area of the State of Israel but contains only about 10% of its population. Thus while the population density of the central part of Israel is about 3,000 persons/km^2, that of the Negev is les than 100. Taking into account that Israel's population has a relatively high standard of living (annual GNP nearly US$ 18,000 per capita) and is mostly urban, about 92% living in cities, while the rest live in rural communities and only about 3% of the total population is engaged in agricultural production. Yet, at present, most of the country's demand for agricultural products is supplied by the local agricultural sector due to the most advanced agro-technologies, including fertigation, high efficiency and the continuing introduction of new crops, By the year 2025 the population of Israel may reach 9 million people,[4] and will need more agricultural products, yet the answer to this demand will not depend mainly on increasing the agricultural population and area of field crops and plantations, but by extending the areas of highly intensive agriculture, namely greenhouses, or plastic covered fields. Yet there is still a need to enlarge the areas of various plantations. In this connection it should be taken into consideration that as the limiting factor is water, the increase in demand of this resource for agriculture can either come from desalinized brackish water,[5] recycled water, whether reclaimed sewage or urban runoff. Fossil water from the limestone and sandstone aquifers is an additional source.

Taking all these future scenarios into consideration it can be foreseen that the main rationale of the development plan of the Negev Desert is to create environmental conditions which will encourage people to move and settle in this region, either in urban or rural centers creating at the same time the infra structure to enable the construction of agricultural projects, either sophisticated greenhouses, or plastic covered farms or plantations.

10.6 The "Artificial Oasis Habitat"

This urban-rural habitat will aim to fulfill, on one hand, the general aspiration of the population to live in urban centers, but in the same time will strive to alleviate the

---

[4]Israel's Central Bureau of Statistics, Jerusalem, December 2004, *Projections of Israel's Population until 2025.*

[5]The cost of desalinized seawater is too high for ordinary agricultural products.

harsh climate conditions of the desert surroundings to allow the comfortable life of a population looking for a high standard of living. At the same time it will enable the establishment of settlements in which high-tech industry as well as advanced agro-technology industries, can coexist. As a matter of fact, these centers will be artificial oases, which will take advantage of most modern advances in technology.

The following suggestions are based on the present author's personal experience in the natural and artificial oases of Iran. These oases are based either on springs or on the water derived from the qanats (chain of wells) dug by the local people centuries ago. Additional experience is from living for about 25 years in the campus of the Institute of Desert Research of the Ben Gurion University of the Negev. The water supplying and greening this oasis is pumped from the limestone deep aquifer underlying the Negev.

The Artificial Oasis Habitat will be built around an artificial pond, surrounded by shading trees. In the Negev the water will come from the one of the aquifers underlying this region. The trees will be irrigated by recycled effluents. The design of the pond will include having most of it shaded by its surrounding trees, to avoid evaporation and reduce the heat. The public facilities, education institutes, community hall, shops etc. will be located around the pond. The living quarters will extend along shaded lanes, branching off from the central pond.

10.7 The Arava Rift Valley, A Special Case

The aquifer in most parts of the valley is built of the Quaternary alluvial fill composed of gravel and sands. In the northern Arava there exists an additional aquifer, namely the Hazeva Formation, of Neogene age. This is built mainly of sands. This aquifer is artesian. The salinity of the groundwater in the Arava is different in each sub-basin and depends on the source of recharge and on local salination processes. The range is from 1,000 to 10,000 mg/l TDS. The two main sources of recharge are from the floods infiltrating into the alluvial fans and inflow from the Nubian Sandstone aquifer. The annual quantity of recharge is about 50 million m^3, but this is not certain yet, and is still under investigation, in first approximation, based on isotope analysis one can say that about 50% of the water is of fossil origin. The outflow is to the Dead Sea in the north and the Red Sea in the south. The north-south groundwater divide line is at Qa e Saidin or Gav Haarava. Where the groundwater table approaches the surface, sabkha conditions are formed. This causes the accumulation of salts at the surface due to capillary action and thus to very saline groundwater in the uppermost layers. At greater depths water of better quality, although brackish, and in confined conditions is found. Israel and Jordan share the alluvial aquifer.

References

1. B. Lucke, *Demise of Decapolis*, Ph.D. thesis, Brandenurg, Technical University of Cottbus, Germany (2007).

2. M. Evenari, L. Shannan, and N. Tadmor, *The Negev: The Challenge of a Desert*. Harvard University Press, Cambridge, 345 pp. (1971).
3. U. Avner, Settlement, Agriculture, and Paleoclimate in Uvda Valley, Southern Negev Desert, 6th – 3rd Millennia BC. In: A. Issar and N. Brown (eds.), *Water, Environment and Society in Times of Climate Change*, Springer, Dordrecht, The Netherlands, pp. 147–202 (1998).
4. A. S. Issar and M. Zohar, *Climate Change – Environment and Civilization in the Middle East*, Springer, Berlin, Heidelberg, New York (2004).
5. A. S. Issar and M. Zohar, *Climate Change – Environment and History in the Middle East*, 2nd edition, Springer, Berlin, Heidelberg, New York (2007).
6. U. Avner, Settlement, Agriculture, and Paleoclimate in 'Uvda Valley, Southern Negev Desert, 6th – 3rd Millennia B.C. In: A. Issar and N. Brown (eds.), *Water, Environment and Society in Times of Climate Change*, pp. 147–202, Springer, Dordrecht, The Netherlands (1998).

Chapter 11
Progressive Development in the Marine Environment

Menakhem Ben-Yami and Arie S. Issar

11.1 From Hunting and Gathering to Agriculture

From about 15,000 years ago as the global warming following the Last Glacial Period dried up many of the countries bordering the desert belts, a few human societies started to give up total dependence on hunting and gathering and step by step began to practice domestication of animals and plants. This enabled gradual transition to a sedentary way of life. About ten millennia ago some human societies adopted a way of life almost totally dependant on agriculture and animal exploitation. This brought in its wake changes in the cultural, social and spiritual world of human societies. Hunting tools became more sophisticated and efficient. The bow and arrow were invented during this period. The introduction of these weapons may have backfired, causing the reduction of wild life populations and prompting the transition to sedentary life and dependence on domesticated animals and growing of food.

While the gradual abandonment of hunting and gathering was a consequence of the need to survive due to the warming up of the globe after the Last Glacial, it seems that the present process of warming will bring about a shift from fishing, i.e. hunting in the high seas and gathering of sea fruits to aquaculture. Various investigations show that the rise of the temperatures of the oceans will affect the marine flora and fauna. Moreover the change in the chemistry of the water of the oceans, namely acidification, will impede the ability of organisms to build shells, ocean circulation influencing population dynamics.

11.2 Marine Fisheries: Capture and Aquaculture

During most of the twentieth century, marine fisheries and fish farming developed in parallel as separate industries with little market interaction. Each had traditional

M. Ben-Yami (✉)
Fisheries Adviser, 18 A.Penn St., 69641 Tel Aviv, Israel
e-mail: benyami@actcom.net.il

A.S. Issar (ed.), *Progressive Development*, Environmental Science and Engineering,
DOI 10.1007/978-3-642-10640-8_11, © Springer-Verlag Berlin Heidelberg 2010

consumers. Some preferred farmed carp or tilapia, while others preferred wild marine fish, like cod and salmon. Lovers of *frutta di mare* have had the choice of fished crustaceans, cephalopods, and mollusks gathered and dredged on their beds in the wild, and of fished or farmed shrimp and mollusks. But recent decades, have seen changes on the fish market. Now many consumers buy fresh or smoked salmon, sea-bream, or frozen shrimp, without knowing, and some without caring, whether they were caught in open sea by fishermen, or grown in ponds or floating cages. The same goes for an additional 15, or so, species of marine finfish whose cultivation is just beginning or already expanding, each according to the progress made by researchers and fish farmers.

Capture fisheries and fish farming are inter-related and to a great extent overlap in their ecology, economics and social impacts. This chapter, therefore, discusses both in the context of the new approach of Progressive Development, which recognizes that our whole civilization has been based on modified ecosystems both on land and in water, and that further modifications are required in view of the badly mishandled needs of the expanding human population and the sustained destitution of many.

11.2.1 The Dilemma of Growing Demand

Worldwide, both governments and international and regional organizations define their marine policies as aimed at achieving *sustainability* in aquatic ecosystems and, particularly with respect to aquatic life and fisheries resources. At the same time, it is widely recognized that the present production of fishery products from the wild and from the existing aquaculture will not be able to match the increasing food needs of the expanding human population.

The commonly proposed solution is both to improve the management of exploited fish stocks, which is widely implemented, although with rather debatable results, and to further the development of commercial fish farming – a process, which is already going on at an accelerating pace. There is a continuum between open-access fishing and intensive aquaculture, along which fishing rights and property rights develop from vague to almost absolute [1]. However, such developments represent the focus of an earnest dispute on a global scale. This, because they affect inter-related processes within the domain of marine ecology, of economics and of the involved social systems of coastal communities and other people involved in fishing, fish farming and fish processing and trade.

In 2004, the total world fisheries yield reached almost 142 million mt, of which some 106 million mt represented food fish, produced almost equally by capture fisheries (58%) and aquaculture (42%) [2]. Aquatic foods have high nutritional quality, contributing 20% or more of average per capita animal protein intake for more than 2.8 billion people, mostly in developing countries. Fish is also the world's most widely traded foodstuff and a key source of export earnings for many poorer countries with particular significance for small island states.

11.2.2 Defining Sustainability

Concerning management of wild fisheries, both critics and advocates of the presently dominant policies of the management establishment claim that they favor *sustainable* extraction of fish and other marine organisms from the ecosystem. They all accept the UN's definition of "sustainable development that meets the needs of the present without compromising the needs of future generations to be achieved through "... balance between environmental integrity, social development and economic development ...". According to FAO, sustainability entails the notion of development, which has no negative effect on the environment and on the future of the resource concerned [3].

This definition, however, has become the subject of different and often contradictory interpretations [4]. Sustainable development means at least three different, often incompatible things to the economists, sociologists and environmentalists [5]. At the two poles of the controversy there are the extreme "nature-first" conservation approach on one hand, and a "development and business first", on the other. The former is about maintaining or returning marine ecosystems as close as possible to their "virgin" or pre-industrial state [6]. The latter is about extraction of fishery resource in the most profitable way, now [7]. Gary Sharp, a widely respected veteran marine systems ecologist wrote on the Fishfolk Internet List: "The word 'Sustainability' is a nonsense term, too vaguely defined to actually be meaningful in communicating the issues, the solutions, and coping with *natural variabilities* – that actually dominate all issues – including how over 6 billion people are walking the earth today"...

Sustainable development has been advertised as a methodology that allows "holding the cake of undiminished resources (including fish) and enjoying the eating of it too". But the prevailing "sustainable development" policies rarely examine their objectives in terms of external costs as, for example, the non-renewable fossil fuels that would be required to carry out their sustainable recommendations [5].

11.2.3 Progressive Development and Management of Fisheries

This section introduces another concept namely progressive development and management policies of fisheries. It entails profound changes to the prevailing approach to management of aquatic resources both in the developed and under-developed areas. It questions those interpretations of the "sustainability principle" in the aquatic realm that focus on preserving lakes, seas and oceans in their "natural" state, and on reverting of some regions to their pre-modification status, while neglecting basic needs of many societies and uncounted communities [8, 4].

This concept is about development of aquatic ecosystems (freshwater, brackish and seawater) that puts human beings and their societies and communities first, with minimum damage to the environment, minimum pollution and with great attention to maintaining maximum feasible bio-diversity. On the other hand, progressive development takes into account the present and future achievements of technology and science, and the maturing of political will that will produce solutions to problems of the present, especially those of coastal and marine pollution, overfishing, preservation of important inshore habitats and offshore nature reserves, and marine farming, which is not detrimental to the environment.

In short, for all practical purposes, fisheries management must seek the golden mean: exploiting fisheries resources without depleting them. Fisheries management means managing the process in which *fishing people* exploit in a rational manner *fish-resources* within *fishery ecosystems*. In every fishery those three are bound together, influenced by such external factors, as people's cultures, markets, technology, and logistics on one hand, and fishery-independent natural, biotic and non-biotic trends and fluctuations on the other [9–11].[1]

The prevailing fishery management system is based on stock assessment supposed to be provided by "the best available science". Unfortunately, this science is mostly inadequate, and in some cases utterly fallacious. It uses simplified assumptions, relies on statistics, avoids ecology, and ends up with often dubious appraisals of whether and how much fish stocks are overfished.[2]

According to Beverton [12] only close liaison between biological and physical research can tackle the effect of long-term climate change on fish stocks in an integrated manner. Multi-species and ecosystem research is vital for elucidating these long-term effects, the source of which lies in profound changes in the early life history of species and in basic productivity. The total amount of fish eaten by other fish, marine mammals, and birds is as great as or greater than that eaten by man. Fishing is only one factor and regulation by catch limits is fundamentally flawed, except in the simplest of single species fisheries, and the total allowable catch (TAC) system is both wasteful and ineffective, Presently, some 15 years later, increasing number of fishery scientists and fishermen are coming to agree with Beverton's statement.

Copes [13] suggested abandoning mathematically impeccable naïve models in favor of a realistic, multi-disciplinary approximation of a working fishery, on the understanding that empirical verity should take precedence over theoretical precision. "For policy implementation in a real-life fishery, is it not better to be approximately right than precisely wrong?" asked Prof. Copes.

Requested for precision by their superiors in the management establishment, stock-assessment scientists keep providing ridiculously precise figures for the stock biomass in the sea and, hence, for the catch quotas they are paid to advise on. No accuracy can be achieved by dividing estimate by approximate, adding a guesstimate and multiplying it all by an assumption.

At best, the dominant fishery science produces very approximate results. At worst, it employs inadequate models and feeds them with inadequate, often erroneous data and their results' correspondence to reality is all but incidental.

---

[1] M. Ben-Yami, Observations on Fishery Management of the Faroes. – Internet: www.benyami.org and at Faroese Fishermen's Union site: www.fiskimannafelag.fo, and as: Making sense of sustainable fisheries management (pp. 17–40). In: M. Olsen (phot.), M. Ben-Yami & B. Tyril. Images of Fishermen: the North Atlantic. (GlobalOne Press Ltd. Aberdeen, UK, nais@globalone-press.com) (2006).

[2] Ben-Yami, op. cit. (2006).

11.2.4 Environmentalist Advocacy and Fisheries Management

While environmentalists and fishermen should join in common cause to maintain resource *sustainability* and oppose damage to the environment due to pollution and coastal habitat destruction, too often they see each other as enemies. Their actual common enemies are the dozens of toxic chemicals and excess nutrients that abound in upstream water and air pollution and their joint effect on coastal marine areas. In addition, the most deadly pollution that affects coastal habitats including fish nursery and feeding grounds, comes from industrial sources and, in particular, from the petro-chemical and electro-chemical sectors [14]. Other factors ignored are the historically documented crucial influence of climatic and oceanic processes and fluctuations on fish populations [15, 16]. Species with narrow temperature preference limits, are affected by thermal anomalies that delay or hasten spawning and hatching, and displace spawning and feeding grounds. Survival of larvae and juveniles depends, apart from hydrographic conditions, on availability of the right food at the right place and time as well as on the rate of predation.[3] Changes in river flows raise or reduce salinity in estuaries, deplete and displace fish and aquatic plant species. Climate variability is the key controlling factor in fishing yields for about half of the world's large marine ecosystems, including the East and West Greenland shelves, the Benguela Current off Southwest Africa, the Canary Current off Northwest Africa and the Humboldt Current off the west coast of South America.[4]

Over the past decades, modes of oceanic variability such as the Pacific Decadal Oscillation and the North Atlantic Oscillation have been related to shifts in marine ecosystem structure, species composition, distribution and biogeochemical processes. Thus, since patterns of species abundances shift with the ocean climate, when climatic, biological and oceanographic conditions are just right, fish can respond with an extremely strong year class, or a series of them, and the other way around [17–20, 9].

The quota system which has been introduced to fisheries management, particularly the individual transferable quotas (ITQ), had great theoretical appeal to economists. They embraced ITQ-system quickly and uncritically, and appeared to regard it as a cure-all for fisheries management problems. While the overall benefits of the ITQ-system are highly disputable, it clearly causes an ongoing concentration of fish resources in the hands of a smaller number of larger operators and a resulting displacement of fishermen owner-operators. Thus, the introduction of the ITQs has become also a political tool for redistribution of benefits from marine fishery resources away from the small to the hands of large firms and corporate interests [21–23].

Evidence accumulated over many years shows that there's more than a single viable resource management option. Private, state, and communal property and

---

[3]Sharp et al. (2004).

[4]Ibid.

fishing rights are all potential options and there's no panacea for fisheries management. Fishery laws and regulations should be tailored separately for every fishery and environment, and for the different socio-economic conditions. The tunnel vision of some economists, which disregards external costs and benefits, represents achievement of the highest profits and free marketing of fishing rights and quotas as also the highest social and national benefits. Often, consequences of such approach becoming actual policy have been social and economic devastation of fishing communities and dislocation of fishing people.

11.2.5 Ecosystem-Based Management

In recent years, fishery science started internalizing the notion that management by single species and only by controlling fishing is illogical and contrary to ecological realities. A new approach called "ecosystem-based management" considers a broad spectrum of influences. The first element is the *time factor*. Any aquatic ecosystem is a dynamic, pulsating, and ever-changing macro-organism. Thus, trends and fluctuations that have been occurring in the ecosystem throughout history must be taken into account [24]. The second element embraces all the changes and physical, biological and chemical forces, including upstream and coastal pollution, imposed upon the ecosystem by *climatic variations* and the various *human activities*. Whatever happens with inshore and bottom habitats, and the water, affects the ecosystem's biota (i.e., all living things). The third element is made up of the *relationships* between the various species occupying the ecosystem at all stages of their life. This runs from bacteria and phyto-plankton up to top predators, with special attention to the managed "target species", physical factors affecting them, their food, prey and predators. Last, but not least is *fishing*, which apart from massive removal of marine organisms from the system, also may influence the size, age and genetic composition of the fished populations [25]. The fishing itself is influenced by the market and the socio-economic context of fishing people and their communities, and of fish consumers, as well as by the industries and technologies involved [24]. No doubt, controlling fish harvests is not enough to ensure sustainable fishery and healthy ecosystems.

"The vastness of linkages between species and critical habitats in a coastal area requires comprehensive management of all its parts" [26]. An appropriate approach should incorporate institutional arrangements and cultural factors to provide for better analysis and prediction [27]. The list of the factors other than fishing that affect the size, composition and well-being of populations of marine organisms is extensive. Here are some examples:

- Further expansion of dead anoxic zones, 150 of which were reported in 2003 in bays and semi-enclosed seas, some of which extended up to 27,000 miles^2, may well be a greater peril than overfishing.
- A slow killing mycobacteriosis epidemic affected the condition and size of the striped bass population in Maryland and Virginia and in the heavily polluted Chesapeake Bay.

- The word's seabirds consume 70 million MT of food, according to some estimates, as compared to the 80–90 million MT of global marine landings, and the amount of fish eaten by marine mammals worldwide is several times the worldwide ocean fish harvest.
- Climatic fluctuations and separate events cause boom-and-bust sequences among fish populations and other marine organisms. For example, studies show that the main factors critical for salmon abundance are distance to and size of oceanic areas with temperature and food critical to the survival of juveniles, such as the abundance of krill, which is their main food.
- Pollution of rivers and estuaries and longshore development that destroys inshore habitats, affect the reproduction of many fish species whose spawning and nursery areas are in inshore waters.

In most coastal areas, rehabilitation of marine habitats and restoration or protection of essential upstream environments and coastal hydrology functions are essential to future fisheries sustainability. The fundamental message is that for such sustainability we must restructure general coastal management. Environmentalist NGOs and fishing interests must try to secure sustainability of habitats balancing between the needs of the people and the conservation of living resources, and both sides need to be prepared to compromise [28]. Furthermore we have to accept that the dynamics of the system is so complex that even the best existing models are of limited use for forecasting the outcomes of actual management actions.

In this general context, "ecosystem-based fishery management" must address particular elements [28], those that require intervention must be defined and managed respectively, while the others, including those that are beyond the management's powers, must be taken into account. Flexibility and tailoring management for each specific fishery and area should become the rule. There is no rational way to impose a policy that doesn't take all the above into account, or a single "common" policy for different fisheries, ecosystems and fishing cultures. We have to analyze separately each fishery ecosystem to see, for example, if TACs are the right methodology or whether other options such as effort control should be considered [29].

While ecosystem-based management is coming of age, institutional inertia and obstinacy, human conservatism, fear of the unknown and need to admit own misconstruction of the realities of the coastal, marine and living resources, impedes its application [30]. The notion of positive development stems from the recognition that, climatic conditions permitting, sustainability of both marine capture fisheries and marine aquaculture can be achieved only by integrated and balanced approach to ecosystem management involving upstream and downstream influences, fisheries, and any other human activities that affect coastal and offshore habitats.

References

1. J. L. Anderson, Aquaculture and the future: Why fisheries economists should care. *Marine Resource Economics* 17, 133–151 (2002).
2. FAO. The state of world fisheries and aquaculture FAO, Rome (2006).

3. J. F. Caddy and R. C. Griffiths, Living marine resources and their sustainable development – Some environmental and institutional perspectives. *FAO Fisheries Technical Paper*, (T353). FAO, Rome (1995).
4. M. Ben-Yami, The meaning of sustainability. Part 1, *World Fishing* 2006 5, 6; Part 2, *World Fishing* 2006 6, 6.
5. G. D. Sharp and C. A. S. Hall, Neoclassical economics and fisheries. In: G. Leclerc and C. A. S. Hall (eds.), *Making World Development Work: Scientific Alternatives to Neoclassical Economic Theory*. University of New Mexico Press, Albuquerque (2007).
6. M. Ben-Yami, Ecological and socioeconomic aspects of the expansion of Nile Perch in Lake Victoria. In: R. M. Meyer, et al. (eds.), *Fisheries Resource Utilization and Policy*. Proceedings of the World Fisheries Congress. Theme 2. Oxford & IBH Publ. Co., New Delhi, pp. 95–110 (1996).
7. R. Goodland, and H. Daly, Environmental sustainability: Universal and non-negotiable. *Ecological Applications* 6(4), 1002–1017 (1996).
8. M. Ben-Yami, Ecosystem management. *World Fishing*, 8, 6 (2005).
9. T. Kawasaki, Why do some fishes have wide fluctuations in their number? – A biological basis of fluctuation from the viewpoint of evolutionary ecology, pp. 1065–1080. In: G. D. Sharp and J. Csirke (eds.), *Proceedings of the Expert Consultation to Examine Changes in Abundance and Species Composition of Neritic Fish Resources*. FAO Fish. Rep. 291(3) (1983).
10. T. Kawasaki, S. Tanaka, Y. Toba, and A. Taniguchi (eds.), *Long-Term Variability of Pelagic Fish Populations and Their Environment*, Pergamon Press, Tokyo (1991).
11. G. D. Sharp, M. Ben-Yami, and J. R. McGoodwin, *Out of Fishermen's Hands: The Evolution of Fishing Cultures*. Center for Climate/Ocean Resources Study, Salinas, CA, USA (Internet: http://sharpgary.org). (2004).
12. R. J. Beverton, The state of fisheries science. In: C. V. Voitglander (ed.), *The State of the World's Fisheries Resources*. Proceedings of the World Fisheries Congress, Plenary Sessions. Oxford & IBH Publishing Co. Pvt. Ltd., New Delhi, India, pp. 25–54 (1994).
13. P. Copes, The extended economics of an innate common-use resource: the fishery. In: A. Eide and T. Vassdal (eds.), *The Ninth Biennial Conference of the International Institute of Fisheries Economics and Trade, Proceedings*, Vol. 2. The Norwegian College of Fisheries Science, University of Tromsø, Tromsø, pp. 889–895 (1998).
14. S. A. Patin, Global pollution and biological resources of the world ocean. In: N. B. Armantrout (ed.), Conditions of the world's aquatic habitat. *Proceedings of the World Fisheries Congress* (May). Theme 1, Athens, Greece (1992).
15. D. H. Cushing, *Climate and Fisheries*, Academic Press, London, 373 pp (1982).
16. L. B. Klyashtorin and A. A. Lyubushin, *Cyclic Climate Changes and Fish Productivity*. VNIRO Publishing, Moscow (2007).
17. R. J. Beamish, Climate and exceptional fish production off the west coast of North America. *Canadian Journal of Fisheries and Aquatic Sciences*, 50, 2270–2291 (1993).
18. R. J. Beamish (ed.), Climate change and northern fish populations. *Canadian Special Publication in Fisheries and Aquatic Sciences*, p. 121 (1995).
19. D. H. Cushing, Climate and Fisheries. Academic Press, London, 373 pp (1982).
20. G. D. Sharp and J. Csirke (eds.), Report of the Expert Consultation to Examine the Changes in Abundance and Species Composition of Neritic Fish Resources, San Jose, Costa Rica, 18–29 April. FAO Fisheries Report Seres 291, Vol. 1 (1983).
21. M. Ben-Yami, Fisheries management: Stealing the common. (A keynote lecture delivered at the St. John's Conference of Harvesters in the North Atlantic Fisheries). *Samudra* 19, 11–18 (1998).
22. M. Ben-Yami, Fisheries management: Hijacked by neoliberal economics. *Samudra* 35, 19–25, July (2003).
23. P. Copes, Problems with ITQs in fisheries management, with tentative comments on relevance for Faroe Islands fisheries. In: S. T. F. Johansen (ed.), *Nordiske fiskersamfund i fremtiden (Nordic fishing communities in the future)*, Vol. 1, Nordic Council of Ministers, Copenhagen, pp. 19–42 (1995).

24. G. D. Sharp, M. Ben-Yami, and J. R. McGoodwin, Out of Fishermen's hands: The evolution of fishing cultures. Center for Climate/Ocean Resources Study, Salinas, CA, USA (Internet: http://sharpgary.org) (2004).
25. T. Laevastu, D. L. Alverson, and R. J. Marasco, *Exploitable Marine Ecosystems: Their Behaviour and Management*, Fishing News Books, Blackwell Science, Oxford, UK (1996).
26. J. F. Caddy, and G. D. Sharp, An biological framework for marine fisheries investigations. *FAO Fisheries Technical Paper* (382). FAO, Rome (1986).
27. D. Feeny, B. Fikret, B. J. McCay, and J. M. Acheson, The Tragedy of the commons: Twenty-two years later. *Human Ecology*, 18(1), 1–19 (1990).
28. S. Carpenter, W. Brock, and P. Hanson, Ecological and social dynamics in simple models of ecosystem management. *Conservation Ecology*, 3(2):4; online: www.consecolorg/vol3/iss2/art4 (1999).
29. M. Ben-Yami, Ecosystem management. *World Fishing*, 8, 6 (2005).
30. M. Ben-Yami, Comment by Menakhem Ben-Yami on "XXI-century alternative to XX-century peer review", *Real-world Economic Review*, 47, 258, 3 October http://www.paecon.net/PAEReview/issue47/CommentsIettoGillies47pdf258 (2008).

Chapter 12
Progressive Development of New Marine Environments for the Production of Marine Vegetation

Amir Neori and Arie S. Issar

The rise of sea level, a consequence of global warming, will devastate the economy of most countries with sea coasts. Estimates of the rate and extent in the rise of sea level, by thermal water expansion and by melt water from glaciers in Greenland and Antarctica, differ among scientists. The Intergovernmental Panel on Climate Change (IPCC) (February 2007) [1] foresees sea-level rise of 0.2–0.6 m by 2100, assuming no rapid melting of glaciers. A major melting in Greenland and West Antarctica will add 7 m to the level of the sea. There are worrisome signs that the process has begun. The eventual disintegration of the West Antarctic ice sheet will add another 7 m to sea level.

According to David Wheeler from the Center for Global Development, the investigation of the World Bank research team that used the latest digital maps, the rapid sea-level rise will threaten millions with inundation. Other papers document signs of unexpectedly rapid change in the Greenland and West Antarctic ice sheets. The possibility of a 3-m sea-level rise in this century is no longer beyond the bounds of informed discussion in the literature. Such a rise is especially significant for countries with economically important and fertile delta region, like Bangladesh, Egypt and Vietnam. A 1-m sea-level rise will inundate more than 15% of Bangladesh, displacing more than 13 million people and cutting into the crucial rice crop. A 3-m sea-level rise will drown Egypt's Nile Delta, and a mere 1-meter rise will inundate much of its fertile land. In Vietnam, the high-risk "red zone", less than 5 m above current sea level, holds 38% of the country's population, 36% of its GDP (gross domestic product) and 87% of its wetlands. As the sea rises many low-lying regions, and entire island nations, such as the Maldives, will disappear under the waves.

Global warming results, among other factors, from the high content in the atmosphere of greenhouse gases (GHG), particularly CO_2. But the rising level of this gas has also its positive aspects, being the main nutrient of photosynthetic plants

A. Neori (✉)
National Center for Mariculture, Israel Oceanographic & Limnological Research Ltd., Eilat 88112, Jerusalem, Israel
e-mail: aneori@gmail.com

A.S. Issar (ed.), *Progressive Development*, Environmental Science and Engineering,
DOI 10.1007/978-3-642-10640-8_12,

and cyanobacteria. As they up take CO_2, plants reduce its atmospheric level when the produced biomass is maintained, in the form of wood, peat and other organic deposits. In fact, fossil fuels are ancient deposits of photosynthetic organisms. Among the plant types that respond most favorably to rising atmospheric CO_2 levels are "C-3 plants", whose primary photosynthetic product is made of three carbon atoms per molecule. At normal atmospheric levels of CO_2, up to half of the photosynthate in C3 plants is returned to the atmosphere by photo-respiration, which occurs simultaneously with photosynthesis in sunlight. High levels of CO_2 in the air reduce photo-respiration, causing photosynthesis to be more efficient. Many algae similarly respond favorably to elevated levels of dissolved CO_2.

Due to the rather limited vertical water turnover in the euphotic zone (the top water layer, receiving enough sunlight to support photosynthesis), photosynthesis in the open ocean is limited primarily by supply of nutrients (mainly nitrogen and phosphorus). Therefore, elevation in atmospheric CO_2 levels is likely to increase photosynthesis and primary production mostly in the more turbulent and nutrient-enriched marine coastal waters, in some of the lakes and in artificial water bodies.

The expected higher temperature of seawater especially in the equatorial and mid-latitudes may promote the growth of phyto-plankton, nutrients permitting. But higher temperature in algae-rich and aquatic plant-rich waters may enhance, on the one hand, blooms of harmful algae (HAB's) and on the other hand, respiration and with it anoxia. But higher water temperatures may also reduce photosynthesis, by reducing water density and thus strengthening the resistance of the warmer top layer of the sea to vertical mixing of deep nutrient-rich water.

A major change in ocean circulation may impact significantly the temperature regimes in oceans and in inner seas, like the Caribbean, Baltic, Mediterranean etc. The character, potency and impact of such changes is still under investigation.

Another positive outcome of global warming for oceanic algae in the mid-latitudes, especially where precipitation regimes are governed by westerly rain-storms, is increased light levels because of the decreased cloud cover. Where nutrients abound at the surface, the result can be higher algal primary production.

Of all the effects of global warming, the single one that seems inevitable and that leads to an unthinkable global catastrophe is the rising sea-levels. Thanks to the advancement of science, however, this can be forecasted, though still not in detail, and prepared for. Engineering devices, especially dikes like those of the Delta Works of The Netherlands, may be of help to prevent flooding in certain regions, while for others there may be a partial help.

It is vital not to delay it to the last moment, but to begin preparations and planning of a protection program as soon as possible in the regions most susceptible to inundation. Pilot projects to evaluate the leading preliminary conceptual models and protection approaches should begin immediately in the most endangered regions. Results will help the planners in the selection of the conceptual model and the protection approach most suitable for the region and proceed with detailed design.

Here we propose a concept that can use global warming to bring the sea-rise protection works into good use, in CO_2 capture and wastewater treatment.

The proposition consists of positioning the dikes or seawalls several km beyond the current coastlines on the continental shelves. The enclosed sea areas will be converted into huge coastal lagoons, in which intensive growing of algae and aquatic plants will be carried out. The lagoons will serve as buffer zones between the open sea and the low-lying population and agriculture regions (essential in case of breaches in the seawalls). The large production of marine biomass will promote an infra-structure of businesses and jobs in the lagoons, and a super-structure of marine biomass-dependant food and energy producing projects. Algae and seaweeds produced in the lagoons will become commodities in the human food (sushi), feed (fishmeal replacement), chemical (agar-agar, carrageenan, alginate) and health product industries, and also support aquaculture of herbivorous and omnivorous shellfish, fish and shrimp. Recently use of seaweeds for bio-fuel has been gaining much interest too [2]. Moreover, while algae and seaweeds can benefit from the consequences of global warming, as already discussed, their growth can actually reduce several of those impacts.

The dikes and polders of the Netherlands are the best pilot project for studying the many problems involved once the protection of the regions threatened by future sea rise is considered. Once a decision to stop the sea has been taken, which in most cases will mean following the Dutch model and build a series of dikes, one of the first questions which has to be answered is whether to adopt a defensive strategy and place the dikes along the existing shorelines or try and turn the future calamity into a benefit, build the dikes off-shore and by this open new prospects for production of food as well as new sources of income related to the production of bio-fuel and last but not least creating new sinks for sequestering atmospheric carbon.

Dikes constructed offshore, the exact distance will be decided according to bathymetric as well as marine conditions, will create large shallow coastal lagoons, with relatively quiet, lighted and clear water, highly loaded with carbon dioxide and low in concentration of other nutrients. In many cases wastewater from household, industrial and agriculture sources will be available from the nearby urban and agricultural regions. Such wastewater, when introduced properly, can support high algal growth [3].

Primary production values from untended hypertrophic warm-water lagoons have been reported at over 12.5 ton C ha^{-1}/year. Semi-intensive IMTA systems in Israel and elsewhere have reported yields of nearly 400 tons ha^{-1}/year of *Ulva* sp. (fresh weight), comparable to 20 ton C ha^{-1}/year in two-species combination culture systems, while *Ulva* sp. monoculture production in tanks reached even higher values [4]. Reported yields of *Ulva* sp. in commercial seaweed farms in South Africa lie between those two values [5]. It is anticipated that seaweed production values of 10–20 ton C ha^{-1}/year can be sustained in any Mediterranean lagoon that will be enriched with wastewater and managed for maximal seaweed growth. Farms of aquatic plants, mostly seaweeds but also sea grasses, will assimilate CO_2 and solar energy to turn nutrient-rich effluents into profitable resources.

The form of algae that develops in a lagoon can be influenced by man to a large degree: Maintaining a balanced supply of nutrients will promote eukaryotic algae over cyanobacteria. Enrichment with silicates will promote diatoms, which are a good food for bivalves, over other phytoplankton [6]. Shallow lagoons, where ample light reaches the bottom, or structures that are submerged at shallow depth, will promote seaweed culture, as most of the seaweed culture in the world is done using ropes or rafts, submerged at shallow depths.

The main use of microalgae that is practical is the in situ feeding of bivalves. All other uses involve the very expensive separation of the cells from the water. Seaweeds, on the other hand, are easily (and cheaply) harvested. Seaweeds of different species can be produced and processed at relatively low cost and have many uses [7], which is why seaweed culture has become the largest sector in world mariculture [8]. Thus, the envisaged lagoons can turn into a source of seaweed biomass and edible seafood, while taking up carbon dioxide and waste nutrients and producing oxygen.

A very promising product from seaweed is bio-fuel. Albeit research is still needed to find out which of the species are most efficient in converting atmospheric and marine sources of nutrition into fuel-valuable biomass, scientists claim that algae, grown in either artificial lakes or lagoons, may help in solving the problems of future supply of car fuels [9]. Research is now focusing on the methods by which algae can thrive to a level, which maximize its industrial and ecological possibilities. The benefits can involve wastewater treatment and the reduction of emissions of carbon dioxide, such as coal power-plant flue gas, which contains about 10–30 times as much carbon dioxide as normal air [10].

Drawing some benefit from the negative impact of power plants, which use fossil fuels–oil, gas or coal, is in tune with the general philosophy of the artificial lagoons idea. An additional consideration is that these conventional power plants will continue to be the source of power, and thus will continue to emit their gases especially CO_2. In many countries these plants are built near the sea, from which water is pumped for cooling and for production of steam. The warm water as well as the gas emitted after burning the fossil fuel can thus be added in intensifying the production of seaweeds in the nearby lagoons.

References

1. IPCC – Climate Change, The Physical Science Basis, Working Group I Contribution to the *Fourth Assessment Report of the IPCC Intergovernmental Panel on Climate Change*, Cambridge University Press (2007).
2. Anonymous—Royal Society. Sustainable biofuels: prospects and challenges. Science Policy, The Royal Society, London, 90 pp. Available at http://royalsociety.org/displaypagedoc.asp?id=28914 (2008).
3. J. H. Ryther, J. C. Goldman, J. E. Gifford, J. E. Huguenin, A. S. Wing, J. P. Clarner, L. D. Williams, and B. E. Lapointe, Physical models of integrated waste recycling—marine polyculture systems. *Aquaculture*, 5, 163–177 (1975).
4. A. Neori, T. Chopin, M. Troell, A. H. Buschmann, G. P. Kraemer, C. Halling, M. Shpigel, and C. Yarish, Integrated aquaculture: rationale, evolution and state of the art, emphasizing seaweed biofiltration in modern mariculture. *Aquaculture*, 231, 361–391 (2004).
5. J. J. Bolton, D. V. Robertson-Andersson, D. Shuuluka, and L. Kandjengo, Growing Ulva (Chlorophyta) in integrated systems as a commercial crop for abalone feed in South Africa: A SWOT analysis. *Journal of Applied Phycology*, 21, 575–583 (2009).

6. J. J. Lee, D. Rodriguez, O. Zmora, A. Neori, A. Symons, and M. Shpigel, Nutrient study for the transition from earthen sedimentation ponds in integrated mariculture systems to ones lined with PVC, what needs to be done? *Journal of Applied Phycology*, 16, 341–353 (2004).
7. A. T. Critchley and M. Ohno (eds.), *Seaweed resources of the world*, Japan International Cooperation Agency, Yokosuka, Japan, 429 pp. (1998).
8. FAO, State of World Aquaculture, FAO Fisheries Technical Paper No. 500. Rome, 134 pp. (2006).
9. M. S. Kelly and S. Dworjanyn, The potential of marine biomass for anaerobic biogas production: a feasibility study with recommendations for further research. The Crown Estate, the Scottish Association for Marine Science, Oban, Argyll, Scotland, 103 pp. (2008).
10. University of Virginia, Algae: Biofuel of The Future? *ScienceDaily*, (2008).

Part III
Progressive Development and Strategic Environmental Assessment

Introduction

"Progressive development" calls for renunciation of traditional Bedouin life styles in the desert, replacing them by a culture of irrigation and water harvesting [1, 2]. If this is implemented, it would essentially mean a departure from the old concept of "sustainability". This sustainability aimed to preserve the current resource availability for an indefinite time by keeping a balance between extraction and renovation of natural resources (particularly water in the case of the desert belts). In contrast, "progressive development" intends to exhaust resources, namely fossil water, for a certain time until human development and environmental change shall allow establishing a new balance based on technology. Is it worthwhile to sacrifice the traditional ways of living for this new concept which is characterised by the belief that man can and should improve himself?

Most discussions of environmental policies focus on reducing the over-extraction of resources, and any concept rejecting the goal of sustainability appears heretical. However, a closer look at history makes clear that change is the essence of the environmental past in the desert belts. Sustainability of a certain life style or production method can only be achieved as long as the environmental framework remains stable–but this stability is a perceived feature of northern European environments, not the desert belts. For example, a small reduction of precipitation by 8% in the humid and semi-humid areas of Jordan means up to 75% less rain for the desert belts [3]. If global warming leads to severe drying of arid and semi-arid regions (ASAR), this could have dramatic impacts on traditional land use.

Since the end of the nineteenth century, the policies of most desert countries were characterized by attempts to develop ASAR according to northern European examples, importing European technologies and concepts in order to achieve European wealth. Simultaneously attempts were made to revive and preserve the specific desert heritage and identity. The inherent tension and opposition of these contrasting concepts might be the root of many failures. Is the idea of sustainability, a concept based on European experiences with over-exploitation of resources, suited to deserts? Or could "progressive development" offer an alternative?

References

1. A. Issar, Progressive development in arid environments: Adapting the concept of sustainable development to a changing world. *Hydrogeology Journal* 16, 1229–1231 (2008).
2. A. Issar, Progressive development to sustain food supply in the arid and semi-arid regions (ASAR). *Journal of Arid Environments* 73, 396–397 (2009).
3. B. Lucke, *Demise of the Decapolis. Past and Present Desertification in the Context of Soil Development, Land Use, and Climate*, VDM, Saarbrücken (2008).

Chapter 13
Some Examples of Development in the Desert Belts

Bernhard Lucke, Iourii Nikolskii, Hendrike Helbron, and Dmytro Palekhov

Most semi-arid regions face strong population growth. For example, with the actual growth rate in Jordan, population doubles every 25 years.[1] Many countries in the desert belt experienced rapid modernisation during the last 50 years. Taking Jordan again as an example, modern western life styles are spreading quickly. They co-exist with traditional herding and farming in the rural areas, but urbanization is rapidly expanding.[2] Most Bedouin tribes have settled down. Modern sanitation needs, industry, and irrigation led to exploding water demands. Most of Jordan's streams are now dammed in reservoirs, and most springs are pumped. Nevertheless the demand can be met only by fossil water reserves like the Disi aquifer [1].

The availability of fossil water led to irrigation of the deserts. In the case of the Disi aquifer, which crosses the border between Jordan and Saudi-Arabia, the latter country attempted to become independent from European wheat imports in the 1980s. Large desert farms started irrigating with fossil water. Heavy subsidies enabled Saudi-Arabia to become not only self-sufficient, but the world's 6th largest wheat exporter until 1992 [2]. However, in 2008 the Saudi government decided to stop all subsidies to agriculture in order to save the rapidly shrinking groundwater.[3] The fossil irrigation water might in the future be replaced by sewage and desalinated water, which however, requires a very high energy input.

In north-eastern Jordan, the ancient town of Umm el-Jimal is an impressive ruin located at the rim of the basalt desert. It started as a Nabatean settlement and continued until the end of the Ummayyad period. During its peak in the Roman-Byzantine era, it was a well-developed desert town with a population of 5,000–10,000 [3]. As the annual precipitation is only 75–200 mm [4], Umm el-Jimal depended on an

B. Lucke (✉)
c/o Prof. Dr. Rupert Baeumler, FAU Erlangen-Nuernberg, Institute of Geography, Kochstr. 4/4, 91054 Erlangen, Germany
e-mail: lucke@web.de

[1] CIA World Factbook, in the internet: https://www.cia.gov/library/publications/the-world-factbook/index.html (31-01-2009).

[2] Ibid.

[3] Reuters, 2008. Saudi Arabia scraps wheat growing to save water. In the Internet: http://www.reuters.com/article/latestCrisis/idUSL08699206 (31-01-2009).

A.S. Issar (ed.), *Progressive Development*, Environmental Science and Engineering, DOI 10.1007/978-3-642-10640-8_13, © Springer-Verlag Berlin Heidelberg 2010 excellent water collection system. Water from periodic floods was collected from two nearby wadis and diverted to cisterns, pools, and irrigation systems. Floodwater farming and herding were the basis of survival. Interestingly, the excavators concluded that the flourishing of the town was a local achievement of the Arab tribes and not the result of the involvement of Roman imperial power [5].

When the ruins were discovered in the early twentieth century, the ancient water collection system was basically operational after an inhabitation gap of approximately 1,000 years. During the 1960s, the main basin which once supplied the garrison of a large fort was cleaned and made available for the local population by USAID.[4] However, after the discovery of fossil groundwater reserves during the 1970s, local landlords started to plant large olive groves, thereby destroying the floodwater collection system and its channel network. Today, the quality of the fossil water is rapidly deteriorating while the ancient water collection system is no longer operational. But the plantations continue to expand although it seems likely that the whole irrigated agriculture will soon collapse. As there is currently no replacement for human freshwater needs, it seems likely that the quickly growing modern village of Umm el-Jimal will soon depend on water trucks from other parts of Jordan.

The Syrian Khabur basin near the site of Tell Shekh Hamad presents another example. Tell Shekh Hamad was a major Assyrian city located approximately 80 km south of the border of rain-fed agriculture. A channel carried water parallel to the river and allowed irrigation by periodic flooding of the fields which prevented salinization [6]. The antique city was abandoned in the third century AD, and the area was thereafter inhabited by nomadic tribes. According to the excavators, the alluvial river basin was once covered by a riparian forest, while the high groundwater table enabled an open park forest and steppe vegetation to survive on the upper river terraces. It was concluded that the present bare appearance of the area is due to human over-exploitation of the environment [7].

Since the 1960s, most Bedouin tribes in the Khabur basin have settled and started irrigated farming. The flow of the Khabur was significantly reduced due to upstream dams and pumping of the river with motor pumps. This and the exploitation of brackish groundwater led to salinization of some soils [6]. The salinization problem is still limited as there was no capillary rise of the groundwater table to the surface. The Syrian government plans to construct a new channel which will essentially follow the course of the ancient Assyrian one, and might allow the continuation of irrigated agriculture if the reduces flow and increases sediment load of the Khabur river to allow continued operation of the channel.

The Dakhla Oasis in Egypt is presented as the last example. Here, ancient irrigation by animal-powered water wheels led to the accumulation of wind-blown sediments along the irrigated, vegetated fields, which grew up to five meters high [8]. After the abandonment of agriculture during the Islamic period, these sediments were eroded by wind which created a ragged terrain of more or less isolated sediment mounds. During the 1950s, the discovery of fossil groundwater at depths

[4]Personal communication with Bert de Vries, January (2009).

of up to 1,500 m made extensive irrigation projects possible, which led to the accumulation of water and efflorescent salinization in depressions. Although better regulation of the water flow has prevented further salinization since the late 1960s, the ongoing extension of agriculture into floors of former lakes raises concern. Since the topography of these areas prevents free drainage, it can be expected that salinization will end agricultural activity in these areas in the near future [8].

References

1. H. Amery and A. Wolf, *Water in the Middle East*, University of Texas Press, Austin (2000).
2. FAO. Irrigation in the near east region in figures. Rome, FAO (1997).
3. B. De Vries, Research at Umm el-Jimal, Jordan, 1972–1977. *Biblical Archaeologist*, 42(1), 49–56 (1979).
4. *National Atlas of Jordan Part One: Agriculture and Climatology*. Royal Geographic Centre of Jordan, Amman (1984).
5. B. De Vries, Paradox of power. Between local and imperial at Umm Al-Jimal. In: T. Levy, M. Daviau, R. Younker, and M. Shaer (eds.), *Crossing Jordan. North American Contributions to the Archaeology of Jordan*, Equinox Publishing Ltd., London, pp. 467–472 (2007).
6. H. Kühne, The effects of irrigation on agriculture: Bronze and Iron Age habitation along the Khabur, Eastern Syria. In: S. Bottema, G. Entjes-Nieborg, and W. van Zeist (eds.), *Man's Role in the Shaping of the Eastern Mediterranean Landscape*, Balkema, Rotterdam, pp. 15–30 (1990).
7. W. Frey and H. Kürschner, Die aktuelle und potentielle natürliche Vegetation im Bereich des unteren Habur (Nordost-Syrien). In: H. Kühne (ed.), *Die rezente Umwelt von Tall Seh Hamad und Daten zur Umweltrekonstruktion der assyrischen Stadt Dur-Katlimmu*, Dietrich Reimer Verlag, Berlin (1991).
8. I. Brookes, Anthropogenic irrigation sediments, Dakhla Oasis, Egypt. In: S. Bottema, G. Entjes-Nieborg, and W. van Zeist (eds.), *Man's Role in the Shaping of the Eastern Mediterranean Landscape*, Balkema, Rotterdam, pp. 113–126 (1990).

Chapter 14
Sustainability in the Desert?

Bernhard Lucke, Iourii Nikolskii, Hendrike Helbron, and Dmytro Palekhov

According to the definition of sustainability, the achievements of past civilizations, e.g. the floodwater farming near the site of Umm el-Jimal, were sustainable since they did not rely on non-renewable resources. But they did not allow continuous settlement from antiquity until today. It is often assumed that breaks in settlement history were due to over-exploitation of the environment, e.g. overgrazing and soil erosion [1]. However the authors agree with Issar and Zohar [2] that natural climate variations and not mismanagement determined the fluctuations in settlement history in the desert belts. For the cases of Umm el-Jimal and the Dakhla Oasis, for example, we think that the gaps of the settlement history were caused mainly by drier conditions.

Whichever explanation is preferred, it is clear that ancient land use was not sustainable with regard to survival or to say it more precisely, that a high price had to be paid. Pastoralism was the human answer to environmental change: if the water stays out, nomads move to the water. Especially tribes relying on camels, as probably preserved in the name Umm el-Jimal (the ancient town name is lost), can endure very dry conditions and cross vast areas in search of pasture and water holes. However, a far lower number of individuals can be supported by pastoralism than by farming, and if no water is found by the herds, death is the ultimate "natural" adaptation to variations of the fragile desert environment.

Due to the blessings of modern medicine and technology, today even the most remote desert areas experience a strong growth of population. Many areas, like the vicinity of Umm el-Jimal and the Khabur basin, are characterised by a semi-development where water resources are exploited to varying degrees without a coordinating plan. And some areas, like the desert of Saudi-Arabia and Dakhla Oasis, were developed intensively, but face enormous environmental challenges. It is therefore pointless to propose "sustainable development" of the deserts. Population growth and contact with modern technology initiated a process during which the

B. Lucke (✉)
c/o Prof. Dr. Rupert Baeumler, FAU Erlangen-Nuernberg, Institute of Geography, Kochstr. 4/4, 91054 Erlangen, Germany
e-mail: lucke@web.de

A.S. Issar (ed.), *Progressive Development*, Environmental Science and Engineering, DOI 10.1007/978-3-642-10640-8_14,

traditional carrying capacity of the desert ecosystem has long since been crossed. The situation is very fragile: it seems possible that the renewable water resources will diminish under the impact of global warming, while the non-renewable water resources may soon be exhausted. And so far there is no sign that the tremendous population growth will soon be reduced. The combined pressure of these developments could lead to collapse, triggering mass migrations, wars, famines, and pandemics. The human answer must be adaptation: a fast development of improved technologies and land use strategies before the current system breaks down.

References

1. W. Lowdermilk, *Palestine – Land of Promise*. Victor Gollancz, London (1944).
2. A. Issar and M. Zohar, *Climate Change – Environment and Civilisation in the Middle East*. Springer Verlag, Heidelberg (2004).

Chapter 15
The Potential of Progressive Development

Bernhard Lucke, Iourii Nikolskii, Hendrike Helbron, and Dmytro Palekhov

The overview of developments in the desert makes clear that Progressive Development is already implemented is many areas, although it is not acknowledged as such and a coordinating master plan is missing. International plans would be very desirable, since water is a transboundary issue. For example, exploitation of the Disi aquifer led to conflict between Jordan and Saudi Arabia [1]. Jordan, Syria, Israel and Palestine share the Yarmouk River as the primary freshwater resource, which is a very sensitive political issue.

Adopting Progressive Development as an international policy goal e.g. by the UN would offer a number of advantages:

- The burdens, responsibilities, and co-operations of the concerned countries could be much better coordinated, and the likeliness of water wars would be reduced.
- The livelihood for the tribes living in the desert could be secured, preventing migrations or mass starvations.
- Carbon could be sequestered by increased vegetation, slowing down global warming.
- Desertification could be reversed.
- Over time, technological progress and internal environmental feedbacks might create a new balance between consumption and renovation of natural resources (water, soil and vegetation).

However, if the Progressive Development of the desert belts fails, an enormous amount of water and capital will have been wasted – and some countries would probably have to be considered as "failed states", falling back to pre-industrial levels of development.

In this article, we evaluate the concept of Progressive Development with regard to three basic questions:

B. Lucke (✉)
c/o Prof. Dr. Rupert Baeumler, FAU Erlangen-Nuernberg, Institute of Geography, Kochstr. 4/4, 91054 Erlangen, Germany
e-mail: lucke@web.de

A.S. Issar (ed.), *Progressive Development*, Environmental Science and Engineering,
DOI 10.1007/978-3-642-10640-8_15, © Springer-Verlag Berlin Heidelberg 2010

(1) Can "progressive development" succeed with regard to the physical environment: will the soils stay productive, or will salinization or erosion bring about an early end?
(2) How can Progressive Development best be implemented with regard to the socio-economic framework? What can be learned from earlier failures of irrigation projects?
(3) Which planning procedures can be used to successfully implement and monitor Progressive Development?

Reference

1. H. Amery and A. Wolf, *Water in the Middle East*, University of Texas Press, Austin (2000).

Chapter 16
The Feasibility of Progressive Development

Bernhard Lucke, Iourii Nikolskii, Hendrike Helbron, and Dmytro Palekhov

16.1 Long-Term Effects of Irrigation on Soils and the Environment

It is a deeply entrenched belief that historic land use led to land degradation in the desert belts. Supposedly, overgrazing and farming caused erosion, and irrigation led to salinization. Considering the gaps in settlement history, mismanagement is considered a prime reason for abandonment [1–3]. If this is so, Progressive Development might be doomed to fail, because the soils are already degraded and will inevitably degrade further. Climate variations, however, can also lead to degradation, and might have caused the fluctuations of settlement history. The correct assessment of the past will be most important for evaluating the impact of Progressive Development. This and the geographic law of soil zonality allow predicting how soils and the environment will develop under irrigation and permanent vegetation.

16.2 The Past as Key for the Future

Paleosols and sediments deliver a long-term record of soil development and are a key for dealing with global warming, since they inform us how soils and landscapes changed under past land use and climate variations. In recent years, important progress was achieved regarding paleosols in semi-arid regions of Jordan and the Levant, mostly limestone soils which are often obscured by calcareous material [4–6]. These studies of the past allow better discrimination between human and natural forces, since the paleosol record covers periods with and without agriculture. Gvirtzman and Wieder [7] showed that paleosols are archives that are suited

B. Lucke (✉)
c/o Prof. Dr. Rupert Baeumler, FAU Erlangen-Nuernberg, Institute of Geography, Kochstr. 4/4, 91054 Erlangen, Germany
e-mail: lucke@web.de

A.S. Issar (ed.), *Progressive Development*, Environmental Science and Engineering, DOI 10.1007/978-3-642-10640-8_16,

to document the crossing of thresholds in landscape dynamics, which makes them best suited to investigate desertification in the light of climatic variations.

In a comparative soil study along a climatic and erosional gradient in northern Jordan, Lucke and his team [8] concluded that calcification and not soil erosion is the governing degradation process. The distribution of soils and sediments over Wadi Ziqlab illustrates how temperature and soil type are related to changing fertility. While the mean annual precipitation is stable over the whole course of the wadi, temperatures vary strongly and follow an altitude gradient of 1,300 m over a valley length of 20 km [9]. The soil and vegetation pattern gives a strong impression of increasing aridity with lower altitude: the contrasting character of the landscape is due to temperature, evaporation, and changing soil properties. Considering the water balance and vegetation, the direct impact of temperature and evaporation seems not as relevant as the indirect role of changing soil properties and local microclimatic controls.

Salinity played no role: even in the permanently irrigated Jordan Valley, good drainage and periodic flooding prevented the accumulation of salts. A detailed soil survey in the Jordan Valley showed that results from the sites sampled by Lucke and his team. (2008) are representative for the greatest part of the valley.[1] It can be concluded that neither dry periods nor historical land use caused salinity in the Jordan Valley.

A trend of calcification was noted in all soil profiles. It seems partly due to the accumulation of calcareous material from the soft rocks along the wadi slopes. Taking the colluvia together, they point to increasing deposition of calcareous dust or steadily reduced leaching processes since the end of the Ice Age. Sediments near the site of Abila showed that a perennial stream dried up after the Byzantine period, and the deposition of soil sediments indicates a trend towards greater aridity thereafter [10].

The rise of $CaCO_3$, and the decline of soil development indices and soil magnetics in the upper part of the studied profiles seemed to be connected not with increasing dust deposition, but decreasing leaching [8]. This means that soil development in the area reflects a general tendency of increasing aridity accompanied by short periods of heavy erosion and soil movement, possibly connected with post-glacial warming. In this context, the age of the prevailing red soils in northern Jordan (Terrae Rossae) was estimated by Nowell's group [11] according to artifacts to 125,000 years and older.

According to Fedoroff and Courty [12], cracking clay-rich soils like Terrae Rossae that are no longer completely hydrated come to a standstill in their formation. Consequently allochtonous material, mostly sand or calcareous deposits,

[1] KfW Feasibility study for the re-use of treated wastewater in irrigated agriculture in the Jordan Valley. Semi-detailed soil survey, Vol. 1, Kreditanstalt für Wiederaufbau, Amman (2005).

fills the cracks, leading to an overall decrease of soil maturity. When the cracks are not operating any more, an important soil feature is missing that could drain excess water from heavy rainstorms. Organic matter content and biological activity are reduced under aridity, which has a negative impact on soil fabric and nutrient availability. According to Bens and colleagues [13], a deteriorating soil fabric may cause soils to repel rainwater when dry, leading to very high runoff and an increased erosion risk, and to a decrease of root penetration depth and vegetation cover. Lower rainfall and higher potential evaporation are also connected with reduced leaching and enrichment of calcium carbonate and other bases. According to Imeson and Lavee [14], higher temperatures in the Judean mountains led to an increasing concentration of soluble salts and therefore directly reduced aggregate stability. Combined with the overall diminished vegetation, aridity thus degrades soils directly and indirectly by lowering their structural integrity and aggravating the impact of floods and wind erosion.

Lucke et al. [8] concluded that several phases of landscape change in northern Jordan were recorded by the soils. The available dates suggest that these changes took place since the end of the last Ice Age, and can be described as:

- An overall tendency to reduced soil maturity, either by decreased leaching or increased deposition of calcareous material, and probable connection with diminished precipitation during post-glacial warming.
- The occurrence of at least two extreme erosion phases which interrupted soil development, and led to deposition of new material until landscape stability resumed.
- The landscape seems stable since the introduction of agriculture, and erosion due to man's activities is negligible–but humans may have contributed to rising $CaCO_3$-content in the soils by releasing calcareous dust.

According to the dates which are available so far, the periods connected with heavy erosion may correspond to the Younger Dryas (11000 BC) and the Yarmoukian landslides (~6200 BC) [5, 6]. It seems possible that heavy flooding recurred during historical periods, indicated by colluvia in Byzantine and Mamluk ruins.

There is no evidence that human activity in the Decapolis dominated soil degradation: even the re-calcification seems to have taken place mainly before 6000 BC. While human activity may have contributed locally to deterioration, the visible development pattern can clearly be attributed to climate, and the last occurrence of global warming at the end of the Ice Age seemingly had a detrimental impact. However, from a nutrient point of view the calcareous Rendzinas seem not inferior to the Terrae Rossae (although it has not yet been evaluated to which degree phosphorus is fixed by calcium carbonate). Does this mean that the visible soil degradation had no effect on fertility? As indicated by high pH-values, the chemical milieu of soils is dominated by base accumulation and not leaching, and chalk samples indicate that nutrients can be bound in significant amounts in calcareous sediments.

Due to this, and because of the strong variation of annual precipitation, manure with artificial fertilizer is not recommended in Jordan[2]. However, soils show a general N-deficiency, which is connected with the low levels of organic matter, and can be attributed to the semi-arid climate: the highest C_{org}-content measured in the study area under oak forest numbered no more than 3.6%. Organic matter also plays a key role for water supply, since it increases the water-holding capacity and has a positive impact on soil structure. The importance of nutrient levels seems generally overruled by the soil's water-holding capacity and rainfall.

It can therefore be concluded that irrigation and floodwater farming as proposed by Progressive Development will have a positive impact on soil development. If sufficient drainage, periodic flooding, and careful use of irrigation water ensure that salts do not accumulate, the enhanced vegetation cover will affect soil development positively. Slightly acidic sewage water rich in organic matter would yield the greatest benefits, and could stop or reverse drought-related degradation as outlined by Fedoroff and Courty [12].

References

1. H. Dregne, *Desertification of Arid Lands*. Advances in Desert and Arid Land Technology and Development, Vol. 3, Harwood, London (1983).
2. D. Hillel, *Out of the Earth. Civilization and the Life of the Soil*, The free press, New York (1991).
3. W. Lowdermilk, *Palestine – Land of Promise*. Victor Gollancz, London (1944).
4. C. Cordova, C. Foley, A. Nowell, and M. Bisson, Landforms, sediments, soil development and prehistoric site settings in the Madaba-Dhiban Plateau, Jordan. *Geoarchaeology*, 20(1), 29–56 (2005).
5. L. Maher, The Epipaleolithic in Context: Paleolandscapes and Prehistoric Occupation in Wadi Ziqlab, Northern Jordan. Dissertation, Department of Anthropology, University of Toronto, Toronto (2005).
6. B. Lucke, *Demise of the Decapolis. Past and Present Desertification in the Context of Soil Development, land Use, and Climate*. VDM, Saarbrücken (2008).
7. G. Gvirtzman and M. Wieder, Climate of the last 53,000 years in the eastern Mediterranean, based on soil-sequence stratigraphy in the coastal plain of Israel. *Quaternary Science Reviews*, 20, 1827–1849 (2001).
8. B. Lucke, I. Nikolskii, M. Schmidt, R. Bäumler, N. Nowaczyk, and Z. al-Saad, The impact of drought in the light of changing soil properties. In: J. Sánchez (ed.), *Droughts: Causes, Effects, and Predictions*, Nova Science Publishers, New York, pp. 69–102 (2008).
9. W. Fisher, K. Atkinson, P. Beaumont, A. Coles, and D. Gilchrist-Shaw, *Soil Survey of Wadi Ziqlab*, University of Durham, Durham (1966).

[2]NSM&LUP Hashemite Kingdom of Jordan, Ministry of Agriculture, Hunting Technical Services Ltd., Soil Survey and Land Resources Centre, Cranfield University, UK. National Soil Map and Land Use Project. Ministry of Agriculture, Amman (1993).

10. B. Lucke, M. Shunnaq, A. Shiyab, M. Schmidt, R. Bäumler, and Z. al-Muheisen, *Questioning Transjordan's Historic Desertification: a Critical Review of the Paradigm of 'Empty Lands'*, Levant (unpublished).
11. A. Nowell, M. Bisson, and C. Cordova, Wadi al-Koum survey. *American Journal of Archaeology*, 107, 453–455 (2003).
12. N. Fedoroff and M. Courty, Soil and soil forming processes under increasing aridity. In: A. Singhvi and E. Derbyshire (eds.), *Paleoenvironmental Reconstruction in Arid Lands*, A.A. Balkema, Rotterdam, pp. 73–108 (1999).
13. O. Bens, U. Buczko, S. Sieber, and R. Hüttl, Spatial variability of O layer thickness and humus forms under different pine beech-forest transformation stages in NE Germany. *Journal of Plant Nutrition and Soil Science*, 169(1), 5–15 (2006).
14. A. Imeson and H. Lavee, Soil erosion and climate change: the transect approach and the influence of scale. *Geomorphology*, 23, 219–227 (1998).

Chapter 17
Modelling the Impact of Climate Change and Irrigation by the Geographic Law of Soil Zonality

Iourii Nikolskii

In order to quantify the impact of climate change on soil fertility, it is possible to apply a mathematical simulation of the relationship between some soil properties (f) and mean annual climatic conditions (I) at the present time. The relationship f(I) can be quantitatively based on the geographical law of soil zonality [1–4] which says that the distribution of the world's main soil types correlates with climate. In other words, the relationship f(I) reflects a permanent equilibrium between virgin soil and climate. In the case of a rather slow climatic change, the relationship f(I) (corresponding to geomorphologically homogeneous groups of soils with similar texture and subsoil mineralogy) should always be conserved. Thus, if the dependence f(I) is established for the present time, it can be used to:

- assess the climatic conditions (I) in the past knowing the property or properties (φ) conserved by virgin paleosols,
- predict virgin and agricultural soil properties (f) according to the climatic index (I) and its change during the twenty first century,
- assess mean annual amounts of irrigation water (by irrigation or flooding) per area unit based on properties (φ) conserved by paleosols,
- assess permissible limits of mean annual irrigation water per area (irrigation requirement) in order to prevent indirect soil degradation by unfavorable hydrothermal conditions for soil formation under irrigation.

According to Volobuev [4], the distribution of properties of some main soil types correlates with mean annual values of the net radiation (R) and precipitation (Pr), or with Budyko's radiative index of dryness (I), which is expressed in the following form:

$$I = R/(L\,Pr) \qquad (17.1)$$

I. Nikolskii (✉)
Colegio de Postgraduados, Campus Montecillo, Mexico
e-mail: nikolski@colpos.mx

A.S. Issar (ed.), *Progressive Development*, Environmental Science and Engineering,
DOI 10.1007/978-3-642-10640-8_17, © Springer-Verlag Berlin Heidelberg 2010

where (R) and (Pr) are expressed in Kj/cm^2 and in mm, respectively, and (L) is the latent heat of evaporation equal to 2.51 Kj/cm^2 per mm. The value of Pr in formula (17.1) expresses the annually infiltrated amount of water in the soil profile. Therefore, it is better to use the difference between precipitation (Pr) and surface runoff (S) instead of only precipitation: (Pr – S). The calculations show that the surface water runoff (S) should be considered only in case of sloped surfaces with more than 3% or in flat areas in humid zones with an annual precipitation of more than 2,000 mm [5]. The mean annual values of (I) reflect climatic conditions: $I < 1$ – humid, $1 \leq I < 2$ – semi-humid, $2 \leq I < 3$ – semi-arid and $I > 3$ – arid conditions.

It is possible to establish quantitative relationships between (I) and regional modal values of some soil properties (f), for example: organic matter content, ratio between humic and fulvic forms of organic matter, calcium carbonate content, cation exchange capacity, base saturation, pH, exchangeable bases, available P and K, content of clay, etc. of the surface layer 0–10 or 0–20 cm inside the horizon A (if possible). The relationships f(I) should be established separately for geomorphologically homogeneous groups of virgin soils with similar texture and subsoil mineralogy, and with the same range of surface slopes.

The relationships f(I) can be obtained using digital sets of modern elevation maps, showing land use, hydrology, hydrogeology and soils (scales 1:50,000 and 1:250,000). The procedure consists of following steps:

1. Sites without agriculture, but with similar geomorphologic conditions and with the same range of surface slopes are selected. The absolute altitude of particular sites plays no role for these relationships because of the dependence of the radiative index of dryness (I) on the altitude.
2. The properties (φ) of typical soils for each site are extracted. It is desirable to collect from soil sampling points covering different climatic conditions with a wide range of annual precipitation and temperatures, and diverse vegetation communities. This permits discovering tendencies of change regarding regional modal values of the property (f), depending on the climatic index (I) and taking natural variation of the property (f) into account.
3. The radiative index of dryness (I) is calculated for each soil sampling point by interpolation of nearby weather station data.
4. A statistical treatment is applied to all pairs of data (f) and (I) corresponding to each sampling point. For that, the total range of (I) values is subdivided into several intervals. Within these relatively small intervals, the random variation of soil properties due to (I) should be insignificant. Then the mean arithmetic values (f_{av}) of each analyzed soil property are calculated inside each selected interval of (I) and all particular values (f) are divided by (f_{av}) in each selected interval of (I). This approach allows considering normalized data of soil properties (f/f_{av}) within every particular interval of (I) as statistically homogeneous samples, and then to determine the type of statistical distribution of (f/f_{av}) values.
5. The graphical dependences of absolute modal values of each soil property (f) on the (I) index are developed.

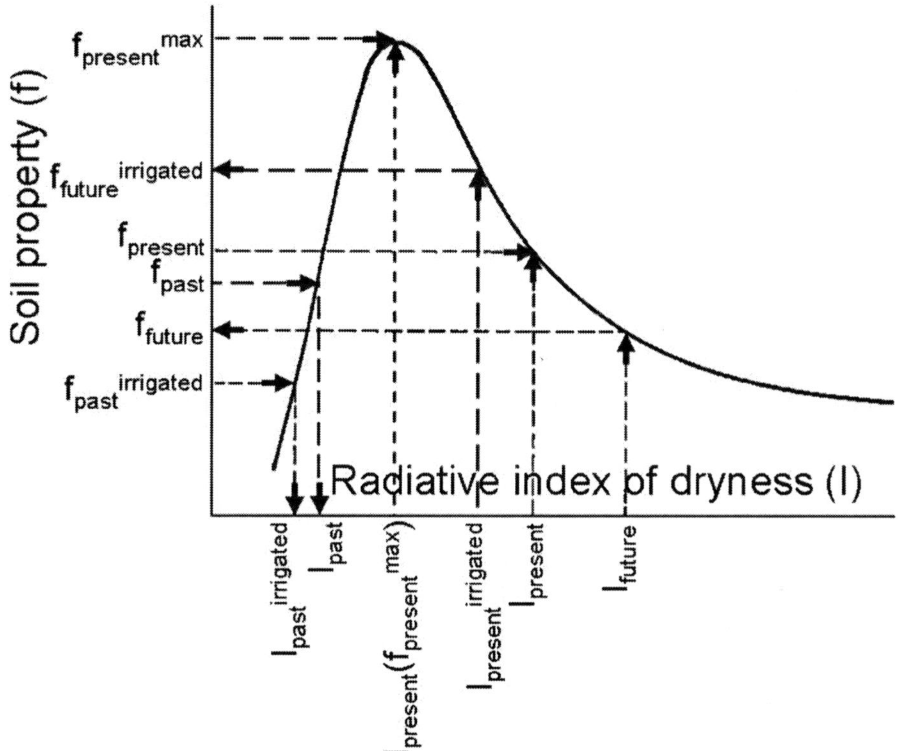

Fig. 17.1 A schematic dependence between a regional modal value of a soil property (f) and the climatic index (I)

The resulting quantitative relationship between a modal regional value of a soil property (f) and the mean annual radiative index of dryness (I) is schematically shown in Fig. 17.1.

If (f_{past}) is a modal value of a paleosol property obtained for a number of samples taken from a climatically homogeneous region or at least from one studied site, it is possible to determine the climatic index (I_{past}) on the graph f(I) established for the present. The index (I_{past}) should reflect mean annual climatic conditions of the past. If the mean annual temperature in the past is known, the mean annual net radiation (R_{past}) can be calculated using empiric relationship between (R) and annual temperature [6, 7].

In case of irrigation, the index ($I^{irrigated}$) should be calculated as:

$$I_{irrigated} = R^{*}/[L(Pr + Ir)] \tag{17.2}$$

where (R^{*}) is mean annual net radiation in the microclimatic condition under irrigation and Ir is mean annual amount of irrigation water application. It is possible to ignore the difference between (R^{*}) and (R) and consider (R^{*}) $\cong$ (R) in case of alluvial flooding or relatively rare irrigation applications.

For the case of past irrigation, the index (I) is expressed by formula (17.2). In case of rainfed agriculture, (I) is calculated by formula (17.1). Using modal values of property (f_{past}) and ($f_{past}^{irrigated}$), it is possible to assess the amount of irrigated water which was applied in the past (Ir_{past}) by the following calculation:

$$Ir_{past} = (I_{past}^{irrigated} - I_{past})R_{past}/L \tag{17.3}$$

The relationship f(I) can be used for predicting an alteration of a soil property (f) due to an expected change of the climatic index (I) during twenty first century. For example, if the index (I) in a studied geographical site is expected to change from ($I_{present}$) to (I_{future}), the modern regional value of the respective soil property should be changed on Fig. 17.1 from $f_{present}$ to f_{future}. The change of a particular soil property [f(local)] of a studied site can be assessed with the following formula:

$$f(local)_{future} = f(local)_{present} + (f_{future} - f_{present}) \tag{17.4}$$

where $f(local)_{present}$ and $f(local)_{future}$ are values of soil property of a studied site at present and future, respectively; $f_{present}$ and f_{future} are regional modal values of the soil property obtained from Graph f(I) for the present ($I_{present}$) and future (I_{future}) climatic indexes, respectively.

It is also possible to use the graph in order to verify whether the ($Ir_{present}$) value points to future soil degradation due to worse hydrothermal conditions. For this it is necessary to remember that the climatic index of irrigated lands ($I_{present}^{irrigated}$) is calculated by formula (17.2). If the climatic index in the site is ($I_{present}$) under rainfed conditions, and under irrigation conditions it is equal to ($I_{present}^{irrigated}$), then it is possible according to Graph f(I) in Fig. 17.1 to observe the future values of irrigated soil property ($f_{future}^{irrigated}$) which correspond to ($I_{present}^{irrigated}$). The irrigation plans can be approved if $(f_{future}^{irrigated})^{3} \geq (f_{present})$; otherwise irrigation will indirectly and negatively impact soil formation.

This can be illustrated with an example. The present dependence on the climatic index (I) of the organic matter contents (OM) in the 0–20 cm soil surface layer of virgin soils of Mexico and the western parts of the former Soviet Union (European part of Russia plus Central Asia) are shown in Fig. 17.2.

All these soils are authomorphous and formed in-situ, located on rather flat lands with surface slopes less than 3%. The sampling points were located at altitudes ranging from 0 to 2,500 m, covered by virgin vegetation and characterized by similar geomorphologic conditions and subsoil properties. The variation of mean annual precipitation and temperature is 200–800 mm and 1–17°C in the Former Soviet Union, and 100–3,000 mm and 6–27°C in Mexico. It is assumed that soil properties come to equilibrium when climate change is rather slow. According to Arnold and his group [10], this applies to organic matter, mineral nutrients contents, and pH. The direct impact of rising CO_2– content on soil organic matter is rather small and was ignored taking into account the estimations made by Bazzaz and Sombroek [11], Reichstein and his group [12], and Knorr and colleagues [13].

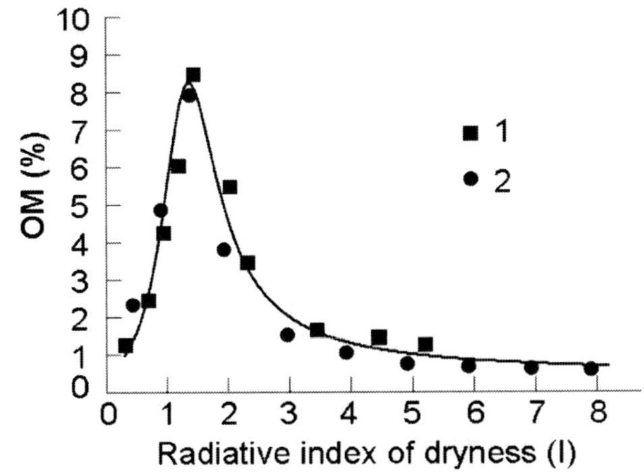

Fig. 17.2 Dependence of organic matter content (OM) on the climatic index (I)

Relationships f(I) were obtained also for cation exchange capacity, base saturation, pH, available P, and K the content of clay in Mexico and for content of water stable aggregates (WSA) and clay, pH and ratio (A_h/A_f) between humic (A_h) and fulvic (A_f) forms of organic matter in the Former Soviet Union. Confidence intervals (with a 95% probability) for changes of all these soil properties (f) numbered $\pm$25% of the modal values of (f). Unfortunately, the dependences [A_h/A_f(I)] and [WSA(I)] were not investigated in Mexico.

The correlation coefficient (r^2) between all modal values (OM) and the approximating curve in Fig. 17.2 is 0.97. As we can see, the relationship OM(I) obtained in both countries with quite different climatic conditions is practically the same. It means that this relationship is typical for such soils. The graphs of pH(I) and clay content(I) relationships published by Aydarov [8] for the European part of Russia and Central Asia, and by Nikolskii and colleagues [5] and Castillo-Alvarez et al. [9] for Mexico, have also similar forms for each property.

The geometric form of the OM(I) graph gives us an idea about the "susceptibility" of this property to climate change. The organic matter content varies most noticeably in the interval $0 < I < 3$. This means that the soils which are more "susceptible" to climate change are located in zones where the dryness index lies within $0 < I < 3$, i.e., the humid tropical zone ($I \leq 1$), the semihumid zone ($1 < I \leq 2$), and the semi-arid zone ($2 < I \leq 3$). The potential impact of climate change may obviously be estimated mathematically as an absolute value of the partial derivative [δ(OM)/δI], meaning the (OM) variation per unit change of (I).

The dependencies of regional modal values of organic matter content and some other virgin soil properties on the climatic index (I) established for the present time can be used for predicting their alteration due to climate change during the twenty first century. It is useful to apply the integral index of soil fertility (F) to assess agricultural soil quality alteration in general. The regional integral index of soil fertility (F) can be expressed in a dimensionless form as suggested by Pegov and Khomyakov [14]:

$$F = 0.46F = 0.46\frac{OM}{OM_{max}} + 0.28\sqrt{\frac{P}{P_{max}} * \frac{K}{K_{max}}} + 0.26e^{-\left(\frac{pH-6}{2}\right)^2} \quad (17.5)$$

where OM, P, and K are organic matter, phosphorus and potassium content, while OM_{max}, P_{max}, and K_{max} are the maximal values observed in the investigation area. The values of F are varied from 0 (for completely degraded soils) to 1 (for the maximal possible fertility in the considered area).

In order to predict the integral fertility index (F) alteration due to climate change, the graphs $F_{present}(I_{present})$ should be established separately for virgin and agricultural soils of rainfed and irrigated lands. If the difference between these graphs is significant, the index of soil fertility (F) can be calculated as:

$$F^{agric}_{future} = F^{agric}_{present} + (F^{max,agric}_{present} / F^{max,nat}_{present})(F^{nat}_{future} - F^{nat}_{present}) \quad (17.6)$$

where $F^{agric}_{present}$ and F^{agric}_{future} are local integral fertility indexes of agricultural soils at present and in the future; $F^{nat}_{present}$ and F^{nat}_{future} are regional integral fertility indexes for virgin soils corresponding on their Graph F(I) to climatic indexes for a studied site at present ($I_{present}$) and in the future (I_{future}); and $F^{max,\ agric}_{present}$ and $F^{max,\ nat}_{present}$ are maximal values of regional integral fertility indexes observed on the Graphs F(I).

If there is no significant difference between graphs F(I) for virgin and agricultural soils of rainfed and irrigated lands, or if $F^{nat}_{present}$ @ $F^{agric}_{present}$, the predicted soil fertility index at a certain site F^{local}_{future} will be:

$$F^{local}_{future} = F^{local}_{present} + (F_{future} - F_{present}) \quad (17.7)$$

where $F^{local}_{present}$ is the current soil fertility index.

An example for the application of this approach to predict natural soil fertility index (F) changes in different climatic zones of Mexico at the end of the twenty first century is shown in Fig. 17.3.

The expected soil fertility change varies between a moderate decrease (up to –30%) to a low increase (up to 20%) depending on the climatic region. A moderate decrease (–20 to –30%) is expected in semi-arid zones of Mexico, a low decrease (–10 to –20%) and low increase (10–20%) in the semi–humid zones, and insignificant changes (–10 to 10%) in the arid zones. The reliability of such predictions depends significantly on the reliability of the climatic scenarios.

The integral soil fertility index (I) can be also used to calculate agricultural crop productivity (Y). According to FAO-IIASA [16] crop productivity can be assessed as:

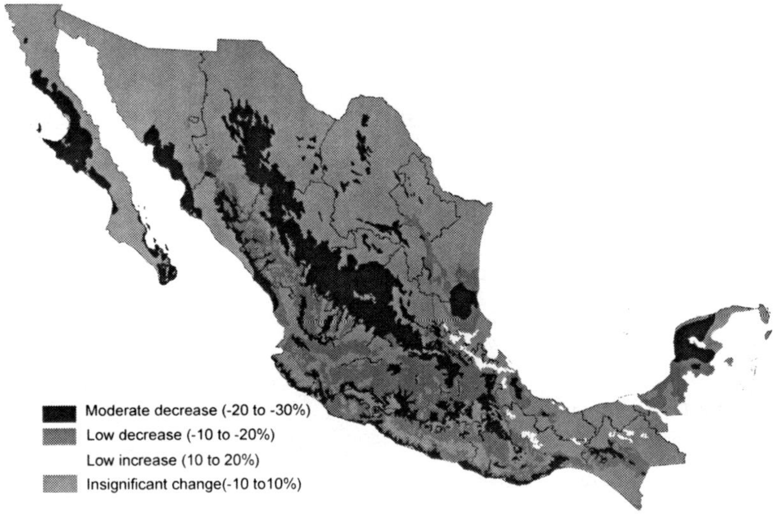

Fig. 17.3 Prediction of the regional integral soil fertility index (F) change in Mexico

$$Y = Y^{max} KF \tag{17.8}$$

where Y_{max} is the maximum possible crop yield (kg ha^{-1}) depending on the predicted monthly values of temperature, global radiation, and biological plant properties; K is a coefficient reflecting soil water availability and depending on predicted monthly values of temperature, global radiation and precipitation; and F is the predicted integral agricultural soil fertility index. K and F are dimensionless and vary between 0 and 1.

A prediction of wheat productivity change in different climatic zones of Mexico under rainfed conditions at the end of twenty first century is shown in the Fig. 17.4. The calculation of crop productivity was conducted using formula (17.8).

The calculations show that the vulnerability of irrigated crops to climate change is *higher* than in rainfed lands, and depends principally on the changing availability of water, especially in arid zones. The vulnerability of rainfed crops depends on changing soil fertility and precipitation due to climate change. Considering soil fertility alteration is very important for assessments of crop vulnerability to climate change: ignoring changes of the fertility index can cause errors up to 50% or more for predictions of crop productivity.

Regarding Progressive Development in arid zones, it is important to consider the indirect impact of irrigation on soil fertility during long-term changes of hydrothermal conditions of soil formation. Irrigation may lead to soil fertility deterioration even if good quality water is used. An example of such phenomena is shown in Fig. 17.5, where relationships between regional modal values of the integral soil fertility index (F) and the radiative index of dryness (I) are shown separately for irrigated and rainfed agricultural and virgin lands (all with surface slopes of less than 3%).

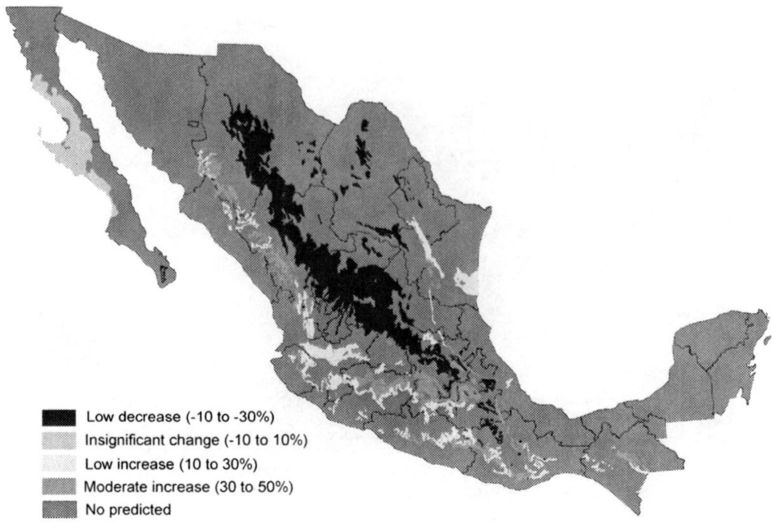

Fig. 17.4 Prediction of wheat productivity change in Mexico

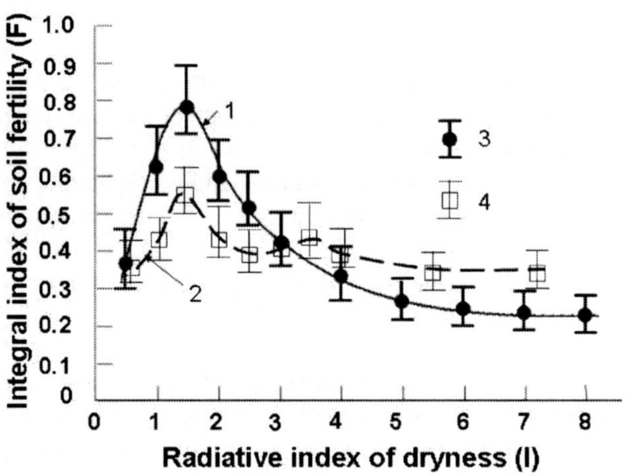

Fig. 17.5 Dependence of the soil fertility index (F) on the climatic index (I) in Mexico

The comparison of modal values of the integral index of soil fertility (F) for virgin and rainfed agricultural soils shows insignificant differences. This is due to extensive land use (without artificial fertilizers and with insignificant manure application) of the majority of rainfed lands in Mexico for maize and wheat production. Therefore only one graph F(I) is shown for virgin and rain fed agricultural soils in this figure.

The impact of irrigation on agricultural soils in arid zone ($I > 3$) in Mexico is in general positive. The fertility index of the irrigated soils is higher than of non-irrigated and virgin soils. However, the irrigation practice in semi-arid ($2 < I \leq 3$)

and semi-humid ($1 < I \leq 2$) zones has negative impacts on agricultural soils, at least in the case of maize and wheat production. These data are representative because they correspond to large areas of irrigated and non-irrigated lands in Mexico (about $3*10^6$ ha for each one). The observed difference between the fertility of irrigated and non-irrigated soils can be explained by:

stronger soil formation processes under irrigation in arid zones and
weaker soil formation processes under irrigation in semi-arid and especially in semi-humid conditions, due to over-irrigation which causes excessive leaching of organic matter and mineral nutrients from soils.

The difference between the fertility indexes of irrigated and virgin soils can be used in order to justify the mean annual irrigation requirements, considering not only agricultural crop productivity (Ir^{crop}), but also long term soil conservation (Ir^{soil}). Such considerations should be part of any sustainable agricultural concept. Regarding irrigated soil conservation, the irrigation requirement (Ir^{soil}) can be assessed using index ($I^{irrigated}$) described by formula (17.2), and graphically shown in Figs. 17.2, 17.3, and 17.5. In climatic zones with $I > 1.5$ (arid, semi-arid and partially semi-humid conditions), the index ($I^{irrigated}$) should number less than 1.5.

This case can be expressed by the following formula:

$$1.5 \geq R/[L(Pr + Ir^{soil})] \text{or} Ir^{soil} \leq (R/1.5L) - Pr \qquad (17.9)$$

for zones where $I > 1.5$

According to Figs. 17.3 and 17.5, the index ($I^{irrigated}$) in semi-humid and partially humid zones ($I < 1.5$) should not number more than the index (I) corresponding to the natural rainfed conditions. This means that in order to prevent soil degradation, the mean annual irrigation requirement should not be higher than the mean annual surface runoff in agricultural lands (S), and less than the difference between potential evaporation (E_0) and precipitation without surface runoff (Pr – S):

$$Ir^{soil} \leq S \text{and} Ir^{soil} \leq (E_0 + S - Pr) \qquad (17.10)$$

for zones where $I < 1.5$

A similar approach was proposed earlier [17].

An example for the determination of (Ir^{soil}), values for some irrigated districts of Mexico are presented in Table 17.1.

In order to prevent indirect negative impacts of irrigation on soil fertility, it is necessary to reduce mean annual irrigation requirements (Ir^{crop}), considering only crop demand. The final decision about the optimal values of mean annual irrigation requirements in practice ($Ir^{practice}$) depends on the climatic conditions. It should be based on the economic justification considering the cost of water and soil resources, and the direct benefits related to agricultural production. With such an approach, the

Table 17.1 Comparison of mean annual irrigation requirements (Ir^{crop}) recommended for soil conservation (Ir^{soil}) in different climatic zones of Mexico

Pr (mm/year)	T (°C)	Ir^{crop} (mm/year)	I (adim)	$I^{irrigated}$ (adim)	Ir^{soil} (mm/year)
528.5	24.6	628	2.7	1.22	409
735	21.3	367	1.6	1.11	250
400.0	19.5	540	2.8	1.19	346
422.8	28.1	798	3.5	1.20	552

Crop – wheat; Pr and T – mean annual precipitation and temperature; I and $I^{irrigated}$ – natural climatic and local microclimatic indexes calculated by formulae (17.1) and (17.2)

irrigation requirement ($Ir^{practice}$) is supposed to be kept between (Ir^{crop}) and (Ir^{soil}): $Ir^{soil} < Ir^{practice} < Ir^{crop}$.

References

1. V. V. Dokuchaev, Russian chernozems (Russkii chernozem). Israel Prog. Sci. Trans., Jerusalem, (1967). Transl. from Russian by N. Kraner. Available from U.S. Department of Commerce, Springfield, VA (1883).
2. A. A. Grigor'ev, Some regularities of geographical zonality, *Izv. Akad. Nauk SSSR, Ser. Geogr.*, 5, 5–17, Russia (in Russian) (1954).
3. H. Jenny, *Factors of Soil Formation. A System of Quantitative Pedology*, McGraw-Hill, New York (1941).
4. V. R. Volobuev, *Introduction into the Energy of Pedogenesis*, Nauka, Russia (in Russian) (1974).
5. Y. Nikolskii, O. Bakhlaeva, A. Contreraz-Benitez, and V. Ordaz-Chaparro, Assessment of changes in soil properties as dependent on hydrothermic conditions of lowlands (by the Example of Mexico). *Eurasian Soil Science*, 35, 1031–1036 (2002).
6. W. D. Sellers, *Physical Climatology*, University of Ontario Press, Chicago, USA (1965).
7. M. A.Tejeda, I. L. Vazquez, and D. A. Rivas, Algoritmos simples para estimar datos mensuales de irradiacion solar y horarios termohigrometricos en la Republica Mexicana. University of Veracruzana, Mexico; Torres, R.E. 1995. Agrometeorologia, Mexico (1999).
8. I. P. Aydarov, *Water, salt and nutrient soil regime management in arid zones*. Agropromizdat, Moscow, Russia (in Russian) (1985).
9. M. Castillo-Alvarez, I. Nikolskii-Gavrilov, C. A. Ortiz-Solorio, H. Vaquera-Huerta, G. Cruz-Bello, E. Mejía-Sáenz, and A. González-Hernández, Alteración de la fertilidad del suelo por el cambio climático y su efecto en la productividad agrícola. *Interciencia*, 32, 368–376 (2007).
10. R. W. Arnold, I. Szabolcs, and V. O. Targulian (eds.), *Global Soil Change*. IIASA, Laxenburg, Austria, 109 pp. (1990).
11. F. Bazzaz and W. Sombroek, *Global Climate Change and Agricultural Production: Direct and Indirect Effects of Changing Hydrological, Pedological and Plant Physiological Processes*, FAO, Italy (1996).
12. W. Knorr, I. C. Prentice, and E. A. Holland, Long-term sensitivity of soil carbon turnover to warming. *Nature*, 433, 298–301 (2005).
13. M. Reichstein, T. Kätterer, O. Andren, P. Ciasis, E. D. Schuize, W. Cramer, D. Papale, and R. Valentini, Does the temperature sensitivity of descomposition vary with soil organic matter quality? *Biogeosciences Discussions*, 2, 738–747 (2005).

14. S. A. Pegov and P. M. Khomyakov, *Simulation of Ecological Systems Alteration*. Gidrometeoizdat, Leningrad, Russia (in Russian) (1991).
15. C. Gay (ed.), *Una visión hacia el siglo XXI. El cambio climático en México*. UNAM, México (2003).
16. FAO-IIASA. Global agroecological zones. Methodology and results of the global agro-ecological zones model. CD-ROM. Versión 1.0. www.fao.org (2000).
17. I. P. Aydarov, *Water, Salt and Nutrient Soil Regime Management in Arid Zones*. Agropromizdat, Moscow, Russia (in Russian) (1985).

Chapter 18
The Socio-Economic Framework: What Can be Learned from Earlier Failures?

Iourii Nikolskii, Bernhard Lucke, Hendrike Helbron, and Dmytro Palekhov

Large-scale irrigation projects are *theoretically* beneficial; however there are examples of outstanding failures which partly led to ecological catastrophes. We think that irrigation per se does not necessarily lead to land degradation. Even in the famous case of the Mesopotamian plains, the idea that ancient Sumerian irrigation caused irreversible salinization [1] is far less evident than often assumed in the public discussion [2].

However, some irrigation projects in arid lands ended disastrously. One of the best-known examples is the former Lake Aral [3]. Irrigated agriculture in the vicinity of the lake basin was applied for many centuries. It was practiced mainly in the areas with rich soils, with a deep water table or with fresh groundwater where the irrigation requirements were minimal. The major part of the irrigated lands was located in river valleys, in deltas, and in foothills close to the mountains. The irrigated plots were small and widely separated from each other. Therefore natural drainage served well enough; artificial drainage, in form of ditches, had been rarely used. The principle irrigated crops were cereals and alfalfa. They occupied more than half of total irrigated area in the Aral Sea basin. Cotton occupied less than 20–30% and rice no more than 5–15% of irrigated area, depending on the years.

During the twentieth century, the population of the Lake Aral basin doubled, and irrigation was intensively developed, leading to almost a tripling of the irrigated area. The basic source of irrigation water is two large rivers, the Amu-Darya and Syr-Darya. Large irrigation systems covering hundreds thousand of hectares were built mainly in the steppe and desert parts of the Lake Aral basin, far from the river beds, where the groundwater table was very deep (more than 30–50 m). Cotton became a main crop, following the government's decision to increase raw cotton production. That led to increased irrigation requirements and to increasing use of mineral fertilizers, pesticides and defoliants [4]. In addition, the area of rice cultivation increased significantly.

I. Nikolskii (✉)
Colegio de Postgraduados, Campus Montecillo, Mexico
e-mail: nikolski@colpos.mx

A.S. Issar (ed.), *Progressive Development*, Environmental Science and Engineering,
DOI 10.1007/978-3-642-10640-8_18, © Springer-Verlag Berlin Heidelberg 2010

The estimated average efficiency on the irrigation systems was very low and did not change appreciably over time. Irrigation was mainly developed by construction of new irrigation systems and less by reconstruction of old ones. About 50% of the irrigation water taken from the rivers was lost because of seepage from canals and deep moisture leaching in the irrigated fields. Even waterlogging took place and the groundwater table sometimes rose to a depth of less than 1.5 m. Due to large amounts of soluble salts in the deep layers of the soils, and as a result of the disturbance of the natural groundwater balance, the worst possible case materialised: saline groundwater rose to the surface.

Later analysis of experimental data showed that the only way to save water and at the same time to prevent soil salinity is to keep the water table at the maximum possible depth (not less than 2.5–3 m), and use only water of good quality. The growing irrigated area and inefficient water use made drainage construction and management of the huge amounts of agricultural wastewater a necessity. Formerly it was supposed that drainage would not only prevent soil salinity and waterlogging, but would desalinate groundwater as well. Then fresh groundwater could have been used for sub-irrigation. Additionally, it was supposed that the repeated use of drainage waters mixed with river waters would increase the effective use of limited water resources. However, it turned out that the salt concentration of drainage waters did not change much. In contrast, it appears that huge amounts of dissolved soil salts, pesticides, and nutrients were injected into the geological water circulation [5]. This contaminated water was used not only for irrigation, but also for human consumption. Therefore the number of severe diseases among the population grew considerably, while the rise of saline groundwater under irrigation systems required increasing amounts of irrigation to prevent soil salinity – despite the newly constructed drainage.

At first, it seemed possible that this problematic situation could be solved with more water, transferring a certain amount of fresh water to the Lake Aral basin. A transfer system was designed: a channel more than 2,000 km long, 12 m deep and 120 m wide should carry Siberian water from the river Ob. But after further analysis, it became clear that the additional water would only make the situation worse, if the present technology of water management remained. About 50% of the existing water resources in the Lake Aral basin were not used, but lost to the seepage in the canals and fields. These losses caused the groundwater table to rise and created drainage water outflows contaminated with toxic agricultural residues and soil salts. Part of the saline drainage water was channelled into closed basins in the desert and evaporates there. This created artificial lakes with a high concentration of pesticides.

To summarise, the results of the Soviet irrigation project in the Lake Aral basin were devastating. The rivers dried up since their flow was entirely used for irrigation. The contamination of irrigated soil and water increased downriver, was severe in the middle parts and most severe in the lowest parts. The level of pesticides in domestic animal meat exceeded the permissible level by 8 times, and in the vegetables by 16 times. The number of serious diseases, such as viral hepatitis, typhoid fever, cancer of the esophagus, increased by 5–30 times. The coastal population at the former Lake Aral lost their jobs, and the endemic flora and fauna in the littoral zone of the former coastal zone disappeared. The annual flooding of the plains does not occur any more, causing strong degradation of formerly rich alluvial soils. Numerous smaller fresh water lakes (some of them used for fishing) dried up. Instead, toxic lakes appeared. Natural pastures for domestic and wild animals disappeared. The amplitude of annual air temperature oscillation has risen by 2–3°C up to a distance of 100 km from the lake. Lake Aral itself has fallen more than 16 m, and is divided into two parts. A dried salty surface appeared on the former lake bottom, covering an area of more than 40,000 km^2 with a salt content of about 100–300 tons/ha in the upper 1 m. About 65 million tons of salty dust are annually blown away from the former lake bottom and carried by the wind for more than 300 km. About 1 ton of salts fall annually on each hectare of the former rivers' deltas. Salty dust clouds rise very high and even reach the mountain glaciers, accelerating their melting. Salty rains with a salt concentration up to 160 mg/l occur in the Lake Aral region. The salt concentration of the lake rose gradually after 1960 and reached 30 g/l at the end of 1980s. According to forecasts, water salinity will increase up to 80–85 g/l even if it is possible to increase the rivers' inflow to 20 km^3/year [6].

Reasons for the ecological catastrophe at Lake Aral include:

- The excessive growth of the irrigated area.
- Inefficient water use.
- Population growth.
- Poor sanitary conditions.

These conditions exist today in most semi-arid and arid areas. If no suitable planning procedures are used, there is a high risk that in the course of the implementation of progressive development large-scale irrigation projects will result in further environmental catastrophes.

References

1. T. Jacobsen and R. M. Adams, Salt and silt in ancient Mesopotamian agriculture. *Science*, 128, 1251–1257 (1958).
2. M. Powell. Salt, seed and yields in Sumerian agriculture. A critique of the theory of progressive salinization. In: D. O. Edzard (ed.), *Zeitschrift für Assyriologie* 75, Berlin, 7–38 (1985).
3. Y. N. Nikolski. Ecological consequences of irrigation. *ICID Journal*, 45, 59–73 (1996).
4. I. Aydarov, Y. Nikolski, V. Jachaturyan and A. Korolkov, La crisis ecologica en la cuenca del Mar Aral por el uso agrícola de los recursos hidraulicos. In: *Agricultura Sostenible: Un Enfoque Ecologico, Socioeconomico y el Desarrollo Tecnologico*. Memorias de /I Simposio y I Reunion Nacional, Guadalajara, Mexico, 79–83 (in Spanish) (1992).
5. Y. N. Nikolskii, Ecological consequences of irrigation. *ICID Journal*, 45, 59–73 (1996).
6. R. M. Razakov, Investigations and programs to improve an ecological situation in the littoral zone of the Aral Sea. *Melioratcia y Vodnoe Hoziaistvo*, 1, 6–8 (in Russian) (1990).

Chapter 19
Strategic Environmental Assessment to Assess and Monitor Sustainable Resource Use in Progressive Development: Potentials and Limitations

Hendrike Helbron and Dmytro Palekhov

Strategic environmental assessment (SEA) is recognised by a vast number of countries worldwide as an important instrument for integrating environmental concerns into development strategies – policies, plans and programmes [1]. SEA is a planning and decision-aiding tool that informs decision-makers about potential effects of development strategies on the environment. In the EU it is legally defined by the EC SEA Directive.[1] As discussed by João et al. [2], SEA has two fundamental objectives. First, SEA must identify feasible alternatives of a strategy and compare them in an assessment context. Second, SEA must improve and not just analyse the development strategy (e.g. land use strategy, spatial plan etc.).

19.1 Improvement of the Progressive Development Strategy

SEA is often referred to in literature as the instrument aiding or promoting sustainable development [3, 4, 1]. The aim of SEA is to ensure that environmental considerations are taken into account equally with social and economic aspects, i.e. prevent risks of overweighting socio-economic issues.[2] Being a flexible planning tool, SEA can be used with equal efficiency for integrating progressive development objectives and aspects of adaptation to climate change into land use strategies. It is

H. Helbron (✉)
Department of Environmental Resources, UNESCO-IHE, PO Box 3015, 2601 DA Delft, The Netherlands
e-mail: hendrike22@yahoo.com

D. Palekhov (✉)
BTU Cottbus, Department of Environmental Planning, PO Box 101344, 03013 Cottbus, Germany
e-mail: dmitry_pal@mail.ru

[1] Directive 2001/42/EC of the European Parliament and of the Council of 27 June 2001 on the assessment of the effects of certain plans and programs on the environment, OJ L 197 of 21.07.2001, p. 30.

[2] K. Arbter. *SUP - Strategische Umweltprüfung für die Planungspraxis der Zukunft: SUP-Erfolgsfaktoren, aktuelle Trends und zukunftsweisende Entwicklungen der SUP. 2.* Neuer Wissenschaftlicher Verlag, Wien. (2007).

A.S. Issar (ed.), *Progressive Development*, Environmental Science and Engineering,
DOI 10.1007/978-3-642-10640-8_19, © Springer-Verlag Berlin Heidelberg 2010

an overall objective to equitably distribute water to the populations in arid regions. In these regions climate change creates new challenges related to water and food security. Therefore SEA should be applied for a transparent mitigation of significant impacts on the natural resources and conflicts with adaptation to climate change, and at the same time, an improvement of the progressive development strategy in arid regions through a strict and systematic integration of environmental concerns.

An example for an optimisation of a strategy for Progressive Development would be alternative water abstractions under consideration of the entire regional water balance, which are sustainable (never is more water abstracted than can regenerate) and linked to precipitation predictions from scientific climate models and monitoring concepts. Moreover improvements of agricultural management techniques with the aim to preserve irrigation water should gain increasing attention in a Progressive Development strategy for arid regions. Limitations are, on the one hand, missing regional environmental quality targets and on the other hand, uncertain predictions of spatial effects of climate change, insufficient regional-specific data and a time delay of implementations. Therefore the use of scenarios and models in SEA can assist to set environmental targets for sustainable water consumption. Furthermore the political willingness and effective implementation of key elements of the SEA process and contents will influence the extent of an optimisation of the strategy.

19.2 SEA Process and Contents

In the literature SEA is commonly defined as a "systematic process" for evaluating the environmental consequences of the proposed development strategies that takes place during their preparation and before their adoption [5, 6]. The systematic character of SEA means that it is an on-going process and not a snapshot at the end of decision-making cycle [7]. Ideally SEA takes place at an early policy-, programme- or plan-making stage, before any action takes place, which allows, for instance, the creation of different development scenarios and the comparison of feasible alternatives. It is a process of communication and involvement during the whole period of drafting the development strategy.

The "classic" (or EIA-based[3]) SEA process consists of a sequence of procedural steps – screening (decision on the necessity of SEA); scoping (determination of the scope of the assessment including aspects of contents, methods, spatial and temporal boundaries etc.); appraisal of alternatives; impact assessment (quantification and evaluation of impacts of all feasible alternatives); preparation of the environmental report (SEA report) and its review; consultations and participation; informed

[3]The SEA procedure based on the process of project-level environmental impact assessment (EIA) is characterised by Fischer (Fischer, T. B. 2007, *The Theory and Practice of Strategic Environmental Assessment Towards a More Systematic Approach*, Earthscan, London) as "structured and rigorous process of predefined steps" and is usually applied for plans and programmes at lower tiers of decision-making. Opposed to this, there is a less formal and more flexible SEA procedure applied for appraisal of policies and legislation at higher tiers of decision-making. See Therivel [4] for description of the "tiering" concept in SEA.

decision-making; follow-up and monitoring. Stoeglehner and Wegerer [8] argued that these steps do not have to follow each other stringently but can be carried out in parallel to some extent, and some may be repeated or combined.

An SEA process has four main results: the most feasible alternative(s) is selected; environmental impacts of the selected alternative(s) are evaluated and described in a written report; the public and other stakeholders are consulted and their opinions are taken into consideration; sound, informed and publicly-accountable decision is made.

Figure 19.1 displays the basic steps of SEA and how they can be integrated into the strategic decision-making process, providing necessary information at all relevant stages.

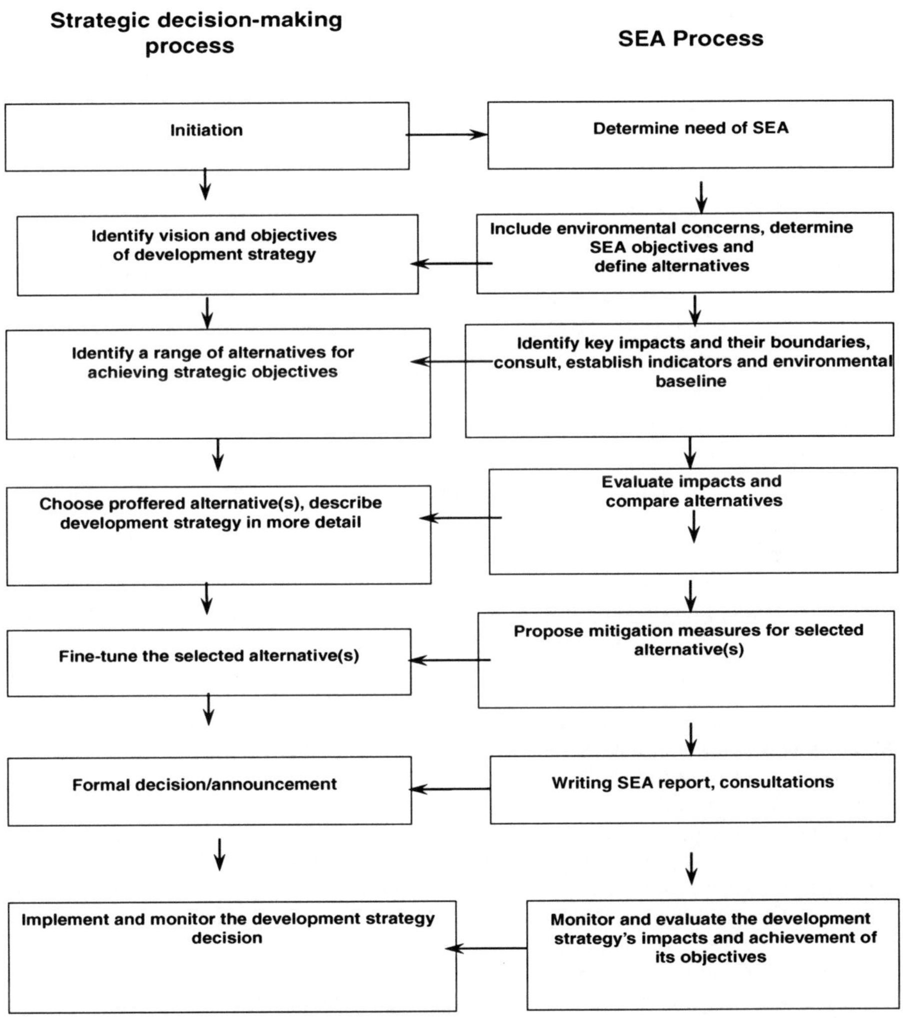

Fig. 19.1 Integration of SEA into strategic decision-making process [3, 4, 7]

19.2.1 Environmental Assessment and Comparison of Feasible Alternatives

The SEA report should include the methods, process and reasons for the consideration and selection of reasonable alternatives (Annex 1 lit. h EC SEA Directive). A transparent and well structured documentation of the no action-alternative and any other alternative proposed and their comparison is required.[4] For instance, in regional land use planning this covers reasons for the designation or rejection of all proposed and assessed site-specific alternatives and any criteria for an alternative prognosis of needs and the selection of conceptual alternatives.

Site-specific regional plan designations and their alternatives are characterised by type and dimension of their predicted impacts which likely accumulate and intensify in their significance in combination with negative effects of climate change.[5] Selection criteria for the conceptual alternative sites can be: minimum areas, minimum space distances, concentration of specific designations in one area; or restriction of specific designations in sensitive natural areas. Conceptual alternatives, for instance, contribute to a reduction of the need to travel, waste production, transportation or energy consumption in the entire region [9, 10].

In Germany typical regional designations with a potential for a comparison of site alternatives in SEA are mining sites, afforestation areas, urban settlement areas, industrial/commercial sites and sites for wind energy production. These designated areas generally are suggested in great numbers and therefore a selection of a certain number can take place and other areas can be held back until the amendment of the revised regional plan. Several designated areas can be compared and the ones with the lowest environmental conflict intensity can be designated.

However, regional planners often do not have the competency to directly decide on whether a plan designation or objective is necessary. In Germany, for instance, they only spatially implement a national or federal targeted area share and production capacity for renewable energies in a region (e.g. the national policy of the German Federal Government is to increase the share of renewable energies at the entire gross power consumption to a minimum of 30% by the year 2020 (Art. 1 (2) EEG[6]). For the Federal state of Saxony the target is 24% by the year 2020 [11]. Both

---

[4]C. Jacoby, SUP in der Raumordnung: Positionen und Praxishinweise von ARL und MKRO. SEA in Regional Planning – Views and Recommendations of the Academy for Spatial Research and Planning (ARL) and the Standing Conference of Ministers Responsible of Spatial Planning (MKRO). UVP report 19(1) 2005, 26–30 (2005).

[5]H. Helbron, Strategic Environmental Assessment in Regional Land Use Planning: Indicator System for the Assessment of Degradation of Natural Resources and Land Uses with Environmental Potential for Adaptation to Global Climate Change (LUCCA). Dissertation at the Faculty of Environmental Sciences and Process Engineering at the Brandenburg University of Technology Cottbus, Germany (2008).

[6]German law on renewable energies – Gesetz zur Neuregelung des Rechts der Erneuerbaren Energien im Strombereich und zur Änderung damit zusammenhängender Vorschriften, 25. October 2008, Bundesgesetzblatt Jahrgang 2008, Teil I Nr. 49, p. 2074.

targets require implementation at the regional level, but often quantitative targets do not exist. Questions of no-action alternatives or whether a certain objective is necessary or adequate require Policy SEA and SEA of the higher state land use plan. But regional planners may be able to decide to fulfil policies to less than 100% in response to regional-specific tendencies, e.g. designation of reserved areas for lignite mining (a non-renewable energy source). Perhaps with the next generation of the regional plan, the need for further mining areas will have changed, as the energy supply by renewable energy sources will have improved in efficiency. Areas with less harm or risk to the environment can be designated in the current regional plan.

In the context of progressive development, alternatives must be considered to minimise the conflict between extraction of fossil groundwater and sustainable water resources, adaptation to climate change and a secure use and allocation of water in the future, i.e. "water security" [12]. A "key message is that, if our global energy habits are the focus for mitigation, the way we use and manage our water must become the focus for adaptation."[7]

The final result of regional decision-making should ideally be sites or strategic options with the least possible negative effect on the environment and global climate and the best compromise between conflicting objectives.

19.3 SEA Application in Practice

A weakness of SEA is that it is still a new mechanism and more experience of its application in practice for different plans and programmes is needed. The role and importance of SEA and adaptation to climate change in the water sector need to be further clarified and the influence of SEA and Integrated Water Resource Management (IWRM) on governance should be strengthened.[8] Earlier failures of irrigation projects due to ineffective assessments, public involvement

---

[7]GWP, Climate Change Adaptation and Integrated Water Resources Management – An Overview. Global Water Partnership, Technical Committee Policy Brief 5. (2007).

[8]R. Hirji and R. Davis, (2009). Strategic environmental assessment: improving water resources governance and decision making, The World Bank, Washington, DC. Water sector board discussion paper series, paper No.12; Slootweg, R., 2009. Integrated Water Resources Management and Strategic Environmental Assessment – joining forces for climate proofing. Perspectives on water and climate change adaptation No. 16. Paper initiated by the Co-operative Programme on Water and Climate (CPWC) and the Netherlands Commission for Environmental Assessment (MER), submitted to the High Level Panel session at the 5th World Water Forum, Istanbul, Internet: http://www.waterandclimate.org/UserFiles/File/PersPap%2016.%20IWRM%20and%20SEA.pdf, last access 11.05.2009; World Bank, 2007. Strategic Environmental Assessment and Integrated Water Resources Management and Development. Economic and Sector Work, Environment Department. http://siteresources.worldbank.org/INTRANETENVIRONMENT/Resources/ESW_SEA_for_IWRM.doc

C. W. Sadoff and M. Muller, Better water resources management – Greater resilience today,more effective adaptation tomorrow. Perspective Paper contributed by the Global Water Partnership (2008).

and implementation by responsible actors demand formal planning regulations integrating a SEA process with a practical assessment method. Experiences of the application of SEA in other sectors and land use planning should be shared, but it is clear, that each SEA is unique and is implemented in a specific context. The application of SEA in a German case study, the TransSEA' project on SEA for the regional plan of Upper Lusatia-Lower Silesia in Saxony, serves as an example in this chapter[9] [13].

19.3.1 Communication and Cooperation with All Stakeholders

First of all it was recognized that continuous communication and cooperation with all stakeholders played a crucial role for the consideration and integration of environmental concerns. Particularly, besides formal scoping meetings, informal transboundary dialogue on the assessment of cumulative impacts was essential. These included, for instance, long-term impacts caused by the common use of water to flood former mining pits and create lakes for recreation and nature. Already existing bilateral cooperation strategies for trans-boundary flood management between Czech Republic and Saxony at the Elbe River were consolidated.

With a changing climate new challenges and problems lie ahead and methods and strategies are still missing to prevent and react to environmental changes directly and indirectly caused by climate change such as heavy rain events. The time span of their occurrence and the significance of impacts are still uncertain. A lot of research is needed on the assessment of cumulative and synergic impacts, which will be influenced by changing natural conditions. The complexity of the interrelations of all environmental media in the context of changing systems due to climate change will never be completely understood. For instance in arid regions droughts might occur more frequently and more intensively in the future than have been predicted. Increasing pressure on water allocation and food production may force more and more people to migrate to countries which do not suffer from water deprivation. Therefore the public should be involved in the development of adaptation strategies and with SEA in water management decisions.

SEA can contribute not only to a stronger awareness of the scarcity of resources, but also to a higher evaluation of natural resources within a society that is dependent on the functionality of the natural capacity. It is recommended to systematically involve the public, particularly in decisions of regional area management with long-term consequences on the regional quality of living, human health and well-being. SEA is deemed to be an adequate decision-aiding system, which can be coupled with the administrative securing of public participation in target setting. If the public

[9] H. Helbron, Strategic Environmental Assessment in Regional Land Use Planning: Indicator System for the Assessment of Degradation of Natural Resources and Land Uses with Environmental Potential for Adaptation to Global Climate Change (LUCCA). Dissertation at the Faculty of Environmental Sciences and Process Engineering at the Brandenburg University of Technology Cottbus, Germany (2008).

is well involved in the setting of regional environmental targets, a certain future decrease of the living quality in the region is more likely to be accepted. The realization of lasting sustainable development in both policy and society demands a continuous updating of environmental awareness. In the context of likely negative effects of climate change on future generations, there is, as yet, insufficient awareness. This will probably change as the population endures such direct consequences as droughts. The higher the public awareness of links between land use planning, predicted negative effects of climate change and adaptation measures, the higher the incentive for and willingness of politicians to set national objectives for stricter protection of the land and natural resources. With an SEA process that is accepted by stakeholders and the public, a long-term optimisation of regional land use planning in terms of improved integration of environmental concerns, integrated adaptation measures to climate change and mitigations of adverse effects of regional development will more likely be achieved.

19.3.2 Environmental Quality Objectives and Standards

Environmental quality standards and objectives are necessary in SEA in order to determine assessment criteria. If the implementation of environmental policy at the regional level is insufficient, an important step in SEA can be the formulation of environmental orientation objectives for a defined time horizon (see Table 19.1).

Table 19.1 Examples of environmental orientation objectives, state of environment indicators and assessment thresholds for the protection of soil and water

Environmental orientation objectives	State of environment indicators	Assessment thresholds (for upper Lusatia-Lower Silesia in Saxony)
Decreased soil loss through erosion: Identified areas at risk of organic matter decline[10] Decreased loss of soil on arable land caused by erosion Restricted soil loss on arable land caused by erosion Strict protection of arable land on slopes against soil erosion	Soils to be protected against erosion assessed in water erosion resistance (EfW) and wind erosion resistance (EfA)	Water and wind erosion resistance of: Special importance: high wind and water erosion risk: EfA 4 and 5/EfW 4 and 5 General importance: moderate wind and water erosion risk: EfA 3/EfW 3 Inferior importance: low wind and water erosion risk:

[10] EC, Green Paper on Adapting to Climate Change in Europe – options for EU action. COM (2007) 354 final, Brussels 29 June 2007. http://ec.europa.eu/environment/climat/adaptation/index_en.htm, last accessed 03.06.2009, p. 18.

Table 19.1 (continued)

Environmental orientation objectives	State of environment indicators	Assessment thresholds (for upper Lusatia-Lower Silesia in Saxony)
Increased area share of extensively managed grassland at the total agricultural land in a region Afforested land with priority on soils, which are endangered by erosion		EfA 1 and 2/EfW 1 and 2; erosion protection forest[11]
Freshwater Resources with Potential for Water Storage and Supply Ensured adequate future water supply and demand management through designation of water reservoirs [14] Protected water quantity of surface waters (rivers and lakes) in their catchment areas Safeguarded and maintained groundwater development rates and levels, and the connectivity of regional groundwater aquifers Integrated framework for adaptation measures from agriculture for the removal of agricultural drainage systems and additional reservoirs for water at lower project level [15, 16]	Regional water bodies with future potential for water storage and supply under spatial effects of climate change (first and second category waters) Land uses with future potential of groundwater development Land uses with future potential of a remediation of the water regime (third category waters)	Special importance: low regional reservoir storage vulnerability under climate change conditions: national primary and secondary water bodies with future potential for water storage and supply; regional water bodies of a high catchment size and a high average discharge; regional small, but in the future recharging, water types with a high water quality; areas with a future high groundwater development rate in >200 mm/a (for OL-NS); fresh or groundwater sanitation areas General importance: medium regional reservoir storage vulnerability under climate change conditions: drinking water protection site zone III; regional water bodies of a medium catchment size and a medium average discharge; areas with third category water bodies with future potential for a remediation of the groundwater regime;

[11] LfUG, 2002. Map of erosion risk. Karte der Erosionsgefährdung flächendeckend im Maßstab 1:200.000, Bodenatlas des Freistaates Sachsen, LfUG 2002, Freiberg. Angaben zur Winderosionsgefährdung EfA. Erstellt auf Grundlage der Bodenübersichtskarte BÜK M. 1:200,000.

LfUG, 2004a. Soil function. Bodenfunktion Böden mit besonderen Standorteigenschaften' – bewertet gemäß Bewertungsinstrument Boden. Angaben zur Wassererosionsgefährdung EfW. Datenquelle BKonz – Bodenkonzeptkarte, M. 1:25,000 (LfUG, status 10/2004). Freiberg.

Table 19.1 (continued)

Environmental orientation objectives	State of environment indicators	Assessment thresholds (for upper Lusatia-Lower Silesia in Saxony)
		areas with a future medium groundwater development rate in 101–200 mm/a (for OL-NS)
		Inferior importance: high regional reservoir storage vulnerability under climate change conditions; water bodies without future potential for water supply and storage (e.g. shallow lakes with small catchment area, without afflux or exchange with groundwater in regions affected by a high precipitation decrease); land uses without potential of a remediation of the groundwater regime; areas with a future low groundwater development rate in ≤100 mm/a (for OL-NS)[12]

19.3.3 Indicators and Assessment Criteria

Secondly, the use of an indicator system in SEA proved to be useful, to make key environmental concerns and important environmental components in the region transparent to the public. Indicators and assessment criteria [17] were suggested, which were linked to the impacts of regional plan designations, environmental quality objectives and regional environmental data. The indicators were presented and discussed in scoping meetings. An indicator system allows a systematic assessment process, which can be integrated into conventional regional plan-making. The consequent application of the indicators and generation of conflict maps will

[12] LfUG, 2004b. Map of groundwater development. Karte der Grundwasserneubildung, 1-km Rasterdaten nach Neumann J, Wycisk P (2003) Atlastafel 5.5 Mittlere jährliche Grundwasserneubildung im hydrologischen Atlas von Deutschland; LfUG, 2006. Gasserhaushaltsberechnung mit GEOFEM-2004. Ermittlung der Grundwasserneubildung für die Region Oberlausitz-Niederschlesien mit Berechnungseinheiten von 500 × 500 m. Dresden.

make site-specific environmental conflict intensities of regional plan designations transparent. The comparison of alternative sites and implementation of mitigation measures can directly address conflicts with environmental objectives and adaptation to climate change. A ranking of potentials of land uses for adaptation will assist in setting priorities of regional activities and use of financial resources. Lessons learned from SEA application can also challenge national spatial and environmental policy.

A weakness of the application of an indicator system may be that data gathering is needed, if current environmental data sets are small. Without a comprehensive information system, the application of the quantitative indicators is feeble. The process of data surveys and derivation of assessment thresholds could slow down the SEA process. At the same time the number of indicators and aggregations are rather complex. It needs to be determined in practice whether a simpler qualitative method may be more appropriate than a complex quantitative method.

19.3.4 Compatibility of Scientific Requirements with Practicality of Approach

Thirdly an important experience from TransSEA was the high challenge for the SEA team to achieve a compatibility of scientific requirements with practicality and the implementation of SEA for a specific regional plan and geographic region. Close cooperation and frequent feedback between the SEA team, regional planning body and stakeholders during the SEA process proved valuable. Statements of the stakeholders led to continuous amendments of the assessment method and contents of the draft regional plan. For instance mitigation measures such as the prevention of wind erosion from lignite mining sites through designation of afforestation sites were integrated at an early planning stage. The very ambitious proposed assessment method was modified by the regional planners to be more practicable. For example, two assessment ranges of a conflict classification were merged to simplify the assessment. From a scientific point of view this can be criticised, as a precautionary range was lost, but from the viewpoint of the regional planning body, the excessive effort demanded in assessing the conflict intensity of over 100 alternative mining sites has to be more reasonable and financially feasible. Therefore it is recommended that a certain level of flexibility of the assessment process and method should be kept, in order to allow effective reactions for instance on changing planning cultures, future environmental concerns and new technological developments (e.g. large-scale photovoltaic installations). The actual purpose of SEA at the strategic planning level should be kept in mind: to optimise a plan and to adequately consider environmental concerns. Consequently the process of mitigating negative impacts itself and tiering become more important than each individual site-specific result of an environmental conflict analysis.

19.3.5 Tiered Decision-Support System

Regional planners should stay open-minded for improvements of their planning culture and give impulse for the higher policy level. Regional level should be strengthened. Communication between sectoral and spatial planners should be further enhanced. Conventional designation criteria should be challenged on their conflicts with principles of precaution and adaptation strategies for a changing climate. A tiered decision-support system for a more sustainable land management should be implemented and maintained for SEA in land use planning from the national to the communal level. Such a tiered system should include information from SEA processes passed on from one vertical level to another or from sector to regional plan level. This could apply restrictions, taboo zones, minimum distances, impact factors, and impact zones. Regional planning should document contents of SEA and further requirements, which are tiered down to binding land use planning. A regional framework can be created with orientation values for settlement densities, population development guidance values, intensity of land use, maximum soil sealing level, guidance values for dense building sites or compensation areas. Areas of general importance need to be assessed in more detail at lower planning and project level, before REP designations are decided or authorisation for project development is given (tiering regulations).

In practice the SEA process often does not exploit its above-mentioned strength to the full. Reasons for this can be a weak implementation of environmental policies or laws at regional level, an unclear embedding of SEA into legal procedures or incompetent authorities and environmental experts. As experienced in the TransSEA project a good participatory, multi-stakeholder process is difficult to implement to its full extent, for instance due to constraints in trans-national communication, administrative differences or unclear competencies in responsible authorities.

References

1. B. Dalal-Clayton and B. Sadler, Implementing strategic environmental assessment. In: M. Schmidt, E. João, and E. Albrecht (eds.), *Strategic Environmental Assessment: A Sourcebook and Reference Guide to International Experience*, Earthscan, London. Environmental Protection in the European Union 2. Springer, Berlin (2005).
2. E. João. Key principles of SEA. In: M. Schmidt, E. João, and E. Albrecht (eds.), *Implementing Strategic Environmental Assessment*, Environmental Protection in the European Union 2, Springer, Berlin, pp. 3–14 (2005).
3. R. Therivel and M. R. Partidario, *The Practice of Strategic Environmental Assessment*, Earthscan, London (1996).
4. R. Therivel, *Strategic Environmental Assessment in Action*, Earthscan, London (2004).
5. B. Sadler and R. Verheem. Strategic Environmental Assessment: Status, Challenges and Future Directions, Report 53. Ministry of Housing, Spatial Planning and the Environment, The Hague, Netherlands (1996).
6. M. R. Partidário, Strategic environmental assessment – principles and potential. In: J. Petts (ed.), *Handbook of Environmental Impact Assessment*, Vol. 1, Environmental

Impact Assessment: Process, Methods and Potential. Blackwell Science, Oxford, pp. 60–73 (1999).

7. J. Glasson, R. Therivel, and A. Chadwick, *Introduction to Environmental Impact Assessment*, 3rd Edition, Routledge, London (2005).
8. G. Stoeglehner and G. Wegerer, The SEA-Directive and the SEA-Protocol adopted to spatial planning – similarities and differences. *Environmental Impact Assessment Review*, 26(6), 586–599 (2006).
9. C. Jacoby, *Die Strategische Umweltprüfung (SUP) in der Raumplanung - Instrumente, Methoden und Rechtsgrundlagen für die Bewertung von Standortalternativen in der Stadt- und Regionalplanung*, Erich Schmidt, Berlin (2000).
10. C. Schmidt, M. Gather, C. Knoll, S. Müntz, S. Wurzler, and J. Blank, *Die Strategische Umweltprüfung in der Regionalplanung am Beispiel Nordthüringens*, Forschungsprojekt im Auftrag des Bundesministeriums für Bildung und Forschung. Fachhochschule Erfurt, Fachbereiche Landschaftsarchitektur und Verkehrs- und Transportwesen (2004).
11. LfULG, 2009. Neue Ziele sächsischer Klimaschutzpolitik 2009. Sächsisches Landesamt für Umwelt, Landwirtschaft und Geologie, http://www.umwelt.sachsen.de/umwelt/klima/16602.htm, last access 29.06.2009.
12. C. W. Sadoff and M. Muller, Better water resources management – Greater resilience today, more effective adaptation tomorrow. Perspective Paper contributed by the Global Water Partnership (2008).
13. L. Stratmann, H. Helbron, S. Heiland, and M. Schmidt, Prüfmethodik und Bewertungsmaßstäbe für die SUP in der Regionalplanung. UVP-report 20(5), 229–235 (2006).
14. CCW – Countryside Council for Wales, Environment Agency, English Nature, Levett-Therivel Consultants et al. (eds.), *Strategic Environmental Assessment and Climate Change: Guidance for Practitioners*; May 2004, p. 6 (2004).
15. EA, 2006a. Impacts on UK land-use. Internet: http://www.environment-agency.gov.uk/yourenv/639312/641094/642206/642375/642472/?version=1&lang=_e, last accessed 12.06.2009.
16. EA, 2006b. Threats to water resources. Internet: http://www.environment-agency.gov.uk/yourenv/639312/641094/642206/642375/642736/?lang=_e, last accessed 12.06.2009.
17. H. Helbron and M. Schmidt, Quantitative threshold values for strategic environmental assessment. In: M. Schmidt et al. (eds.), *Standards and Thresholds for Impact Assessment*, Springer, Berlin, pp. 405–418, Table 1 (2008).

Chapter 20
Discussion and Conclusion

Bernhard Lucke, Iourii Nikolskii, Hendrike Helbron, and Dmytro Palekhov

The current discussion of "sustainability" focuses on aspects of conserving the environmental status quo. However, the environmental history of the desert belts is characterized by strong environmental fluctuations, and modern technology started to irreversibly change societies living in arid regions, initiating a growth which long since crossed the limits that could be supported by renewable resources. Due to these conditions, it seems unlikely that "sustainability" in the conservative meaning of preserving the status quo will be feasible in the desert belts. In fact, Progressive Development is already implemented, though not recognised as such and without coordinating master plans.

A key requirement for the success of progressive development is the allocation of water for irrigation of semi-arid and arid areas. This appears dangerous, since it was often assumed that irreversible soil degradation is the inevitable outcome of intensive land use in the desert belts, in particular of large-scale irrigation in arid lands [1]. However, our analysis showed that this is not necessarily the case. If applied wisely, irrigation technology has the potential to improve soils and fertility. The relationship of soil properties, climate, and irrigation can be modelled according to mathematical formulas, which allow predicting the impact of specific land use practises for given climatic scenarios. With such tools, it should be possible to plan progressive development well ahead as long as the necessary water supply can be ensured.

However, it is clear that large-scale irrigation projects involve a high risk. The tripling of the irrigated area around Lake Aral during the twentieth century was the main reason for the ecological crisis in Central Asia [2]. A growing and unwisely planned irrigation project may involve more and more lands with bad drainage, and larger irrigated areas are connected with smaller specific lateral subsurface water outflows. This may cause devastating irrigation problems such as a capillary rise of the groundwater table.

B. Lucke (✉)
c/o Prof. Dr. Rupert Baeumler, FAU Erlangen-Nuernberg, Institute of Geography, Kochstr. 4/4, 91054 Erlangen, Germany
e-mail: lucke@web.de

A.S. Issar (ed.), *Progressive Development*, Environmental Science and Engineering, DOI 10.1007/978-3-642-10640-8_20,

Ecological catastrophes such as Lake Aral were the result of bad planning and unwise land management. Such failures could be prevented if suitable planning tools such as SEA are systematically and inevitably applied. Prerequisites for an effective SEA process are that the purpose of SEA is understood by all stakeholders, a good planning system is in place, clear environmental objectives and legal restrictions exist, a competent authority and independent control body are responsible, the assessment and decision-making process is made transparent, the public and all stakeholders are involved, and a monitoring concept with feedback loops is established.

Results of SEA in Progressive Development will serve decision-makers as guidance. The final result should be a planning option with the least possible negative effect on the environment and the best compromise between conflicting objectives of land use. However, SEA on its own cannot guarantee sustainable land use. Sound planning requires a framework of strict legally-binding standards or exclusion zones, where adverse effects are prohibited by law such as water extraction restrictions linked to climate and nature reserves. However, the setting of strict environmental standards is a political decision, which is based on different interests and concerns, and therefore faces various hindrances or difficulties. It thus can be a long and slow process. A faster recommended action is to first determine clear responsibilities and activities for adaptation measures at all planning levels.

Strategic Environmental Assessment (SEA) for progressive development has an important potential to contribute to weighted and sound development and use of natural resources and land. As significant impacts of progressive development on water, soil and land are expected, it is recommended that no decisions should be made without an effective SEA, which focuses on environmental impacts and is accompanied by a sustainability assessment (SA), which particularly addresses economic and social concerns.

Progressive Development requires an effective SEA including public participation. The quality of the SEA process will very likely influence the acceptance and dimensions of water extraction and irrigation. Only if a proper monitoring concept is in place, can unforeseen and unpredictable significant impacts be recognized and respective reactive measures can be implemented. Weak institutions, poor stakeholder participation and insufficient resources influence the quality of SEA and its input in decision-making. Therefore capacity building and training should accompany the implementation of SEA in land use planning.

We conclude that the potential Progressive Development is not a matter of the physical environment or the available technology, but of planning and the societal framework. It appears that the future of the desert belts depends largely on the successful implementation of planning tools that allow long-term progressive development.

Acknowledgements We are grateful to the Secretaria del Medio Ambiente y Recursos Naturales (SEMARNAT) and the Consejo Nacional de Ciencia y Tecnología (CONACYT) for supporting the Mexican research, and to German Research Foundation (DFG) for funding the research in Jordan (SCHM 2107/2-1).

References

1. D. Hillel, *Out of the Earth. Civilization and the Life of the Soil*, The Free Press, New York (1991).
2. Y. N. Nikolskii, Ecological consequences of irrigation. *ICID Journal*, 45, 59–73 (1996).

Part IV
Appendices

Appendix A
The Project in NE Kenya an Example of a Development Project Based on Groundwater, Following the Conceptual Model of Progressive Development

In the following appendix an extended summary of the report for Northeast Kenya, written in 1992, is presented. Although the principles of the policy of Progressive Development was not yet put in a proper outline, yet its basic philosophy forms its background. No wonder it was opposed by the proponents of the conceptual model of Sustainable Development. As mentioned, reading the news and seeing the TV documentaries about the poor people starving in this region the author cannot keep himself from thinking that if the recommendations brought up in his report, would have been applied, or in other words the philosophy of Progressive Development would not have been rejected by the devotees of Sustainable Development, then at least a few ten thousands (if not more) of these people would have been saved.

A.1 The Development of Perennial Water Resources for Irrigation for the Settling of the Nomad Population in the Arid and Semi-Arid (ASAL) Northeast Kenya

A.1.1 Preface

The series of droughts which has affected Northeast Kenya during the last few decades has caused devastation to the population and had a disastrous impact on the natural environment. This situation may become worse in the future, due to the global climate change and the rate of increase of the population of Kenya at about 4.3% (while in the ASAL provinces it is about 1%). This explosion of population will, sooner or later, result in an overflow of people into the scarcely populated regions. This may happen if a series of humid years occurs. When a series of dry years will follow, as expected, the impact on the indigenous population, as well as on the new immigrants (not to mention the environment), will be hazardous. On the other hand, with the development of the ASAL areas in a planned way, the newly developed resources of these regions will secure the supply of food even during years of drought. A higher level of income will bring about the promotion of other living conditions, such as education, health, etc. Moreover, development will prevent deterioration of the environment in the case of droughts, and may even absorb, in a positive way, the overflow of population from the regions of high potential.

A.S. Issar (ed.), *Progressive Development*, Environmental Science and Engineering, DOI 10.1007/978-3-642-10640-8, © Springer-Verlag Berlin Heidelberg 2010

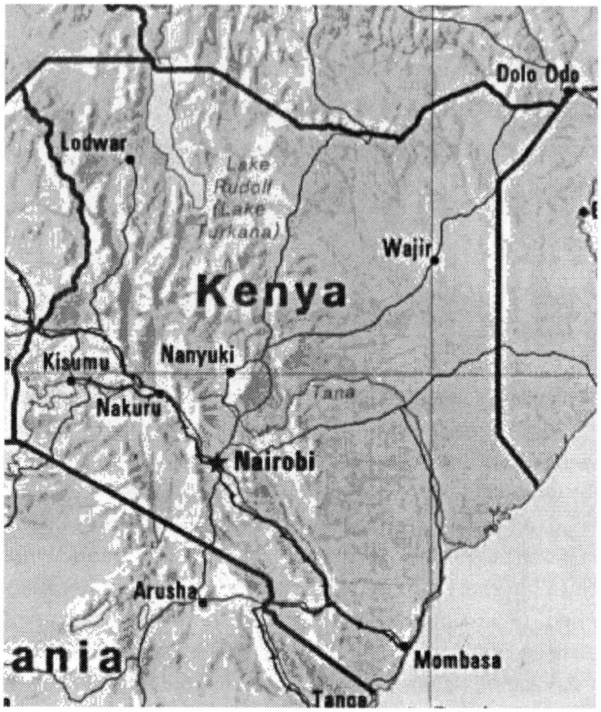

Fig. A.1 Map of North East Kenya

An additional aspect of the development of the ASAL regions is making Kenya a front post and a center for study, experiment, and experience for fighting the hardships brought about by drought and desertification.

The ASAL regions are what they are because the water resources are not sufficient. Yet, it should be remembered that deficiency is only one aspect. The other aspects are the shortness and randomness of the periods with rain events. Throughout the ages this has caused the survival of the population which was wise enough to adopt the pastoral and nomadic way of life. Thus, if any thing is to be done to change the present situation, it must start with ensuring a stable source of water, in space, quantity and time. The sources which fulfill these demands are either the rivers which flow into the ASAL regions from the highlands where high precipitation rates occur, and the groundwater aquifers underlying the regions. The other water resources, such as local floods which may be harvested by special methods of soil tillage, soil embankments and small dams, are important as auxiliary resources but cannot answer the basic demand for survival in years of drought. Yet, any development plan has to be comprehensive and has to cover all resources, physical as well as human. It also has to take into consideration all impairments, such as traditional ways of life, land ownership, etc. Yet, it should be remembered that the will for survival, and even that to secure a better income, will, in most cases, bring about a change in lifestyle. This, if not planned to be procured in a positive way, will take place in a way harmful to the human as well as natural environment. It is, thus, suggested to trigger the change from an insecure mode of life to a safer one by the provision of a stable and safe water supply for human and livestock consumption and, wherever possible, also for irrigation. This does not imply that the other facets of the comprehensive development will be neglected. It does say, however, that priority in time and funds will be given to the prime factors. The development of the other factors will follow suit in the framework of a long-term development master plan.

A.1.2 Summary of Observations

The main observations which brought about the recommendations hereby presented are as follows:

1. Wherever perennial water sources, such as rivers, springs, or groundwater, have been located and made available in ASAL regions, a gradual change from a nomadic to a sedentary, socio-economic style of life has taken place.
2. In most of the provinces visited, groundwater can be located at a depth of several meters to 200 m. In other provinces pumping from perennial rivers, dams on seasonal rivers, and the tapping of springs can also be carried out. Yet, relative to groundwater, the management and distribution of seasonal surface water, needs long-term observations, elaborate planning, and the construction and maintenance of canals.
3. A major obstacle to the optimal development and utilization of groundwater is the dependence on motorized equipment, such as drilling machines and pumps. This is especially true in regions where the water table is deep. This obstacle was, and can be, overcome through the organization of a supply and maintenance system on a regional basis. Such systems are functioning successfully in the province of Nairobi. It can, thus, be concluded that this system can be extended throughout the country. In many regions where the water table is shallow, hand-dug wells are already successfully being used and are pumped by manual, wind, or motorized pumps.

A.1.3 Summary of Recommendations

As stated, the long-term aim of the policy of the project suggested in this report is to promote the potential of the arid and semi-arid zones in order to increase the economic safety and income of the people. This will serve to cope with the recurring hazards of drought and famine, as well as enable these regions to absorb the population overflow from other regions. For this purpose, it is suggested that a policy of development will be adopted, which will encourage the population of the ASAL regions, in the places where the nomadic-pastoral way of life is prevalent, to abandon their traditional livelihood and adopt, through a gradual evolutionary and voluntarily process, a sedentary way of life. At the same time in regions where the population is already sedentary, but still dependent on livestock and dry farming, to gradually adopt the methods of irrigated agriculture. Part of the population will find its income in new urban centers which will supply various types of services to the rural sector.

This transformation will demand the preparation of a long term master plan for development of the ASAL areas. The time period of each plan will be 5-years. The up-staging of the master plan will be from the district to the provincial level. Namely, that after the principles of the overall policy on the national and provincial levels have been agreed upon, it will expand from the district level to the provincial scale. In each province there should be a steering committee in order to guarantee inter-district coordination and promote projects on the provincial level. A list of provinces and districts to be given priority and accordingly allocate the funds available, should be prepared by the government.

It is recommended that this planning be carried out simultaneously with implementation of the projects. It is, thus, essential to guarantee information flow from the planning team to the field team and vice versa. For this reason, the planning team will coordinate the operation stage to its completion. It is expected that variations in the plan will have to be made in accordance with feedback from the field team.

The budget needed for planning is, of course, a function of the decision of the government on the number of districts and on the existing development and availability of data from each district. It is estimated that the average time period needed for preparation of a preliminary plan for each district is approximately 6 months. The detailed plan, which will be done simultaneously with the execution of the plan, will take about 5 years. The planning teams will be comprised of about 5 experts working in the fields of regional planning, farm planning, economics, hydrology and water engineering. The team of experts should comprise foreign experts and local counterparts. The foreign team is expected to stay for about 2 years in order for the work to be continued by the local specialists, with one foreign expert remaining for special assignments.

A.1.4 Development of Water Resources

As previously mentioned, it is recommended to start immediately with the implementation of a general plan to bring about gradual socio-economical change. For this purpose, it is essential to guarantee a reliable perennial water supply for use by man, livestock, and for irrigation. In order to make this feasible throughout the ASAL regions, it is suggested to embark on a long-term, inter-regional campaign in which surface and groundwater resources will be tapped and properly allocated and managed. This means the launching of an inter-regional comprehensive program for exploration, planning and research for water, and new agricultural practices.

All water resources such as rivers, surface and subsurface water will need to be developed. It should be emphasized again and again that as the ASAL areas are susceptible to droughts, groundwater is the substantial source guaranteeing supply when all other water sources have failed. Groundwater is the least influenced by the random character of the climate. The development of the groundwater will need to be carried out in the framework of a comprehensive development plan which takes into consideration the socio-economic as well as environmental aspects within the regions.

It is thus suggested to give priority to groundwater development. It will start with hand-dug shallow wells, equipped with manual and other types of pumps thus upgrading exploitation of the shallow and surface water resources. In parallel comprehensive program for drilling exploration-exploitation wells will be carried out. Location of sites will be done in cooperation with the planning teams.

The drillings of the deep wells will need the organization of teams equipped with drilling machines, casing-pipes, and pumps able to carry out a crash program. After the wells have been drilled and equipped, a special organization in each district will have to be set up in order to maintain the pumps and the wells. The setting up of such organizations is feasible only if a large number of wells are drilled. It is recommended that at least 2,000 wells be drilled within the next 15 years. It is suggested that the drilling will be executed one stage at a time, starting in regions of high priority and being followed by the schedule of the master plans. The annual budget will, thereby, be decided according to the pace of execution.

It should now be mentioned that in order to avoid the already known, and frequently mentioned, phenomena of overgrazing near water sources, a sufficient number of well, shallow as well as deep, should be spread out over each district.

A problem which needs special reference is the impact of development on the environment. It is sufficient to say that the development recommended in this report can also be regarded as an environmental action plan, the aim of which is to change, in a positive way, the gradual degradation of the environment through the insurance of the existence and income of the local population. This will avoid the present trend of over exploitation of the available natural resources, a process which has reached the level of catastrophe in years of extreme drought.

A.1.5 The Impact of Water Resources Development on the Environment

This question has been raised many times during the various discussions regarding strategies for development. Many of the arguments have been against development of groundwater, pointing out the negative impact on the environment which the development of this resource has had in the past. This is because such development has brought about the concentration of livestock around one area, which has caused overgrazing and, thus, deterioration of the natural environment.

The argument against this accusation is not because the remedy was not good enough but because it was not given in the appropriate dosage, and without the other measures needed in order to make the remedy effective. In other words, it is suggested to regard development of groundwater as a measure to be applied on a large scale and backed by other means, such as water harvesting, improvement of grazing sources, gradual change in the socio-economic system, and improvement of the technical know-how of the population.

It is thus suggested to regard groundwater development as the triggering impact for various changes necessary in order to make indigenous societies less vulnerable to the hardships caused by droughts. This can be achieved by the very nature of this resource, where its availability is not influenced by the random supply of water by rains. Thus, when surface reservoirs remain empty, or do not retain enough water to cover the demand during the dry season, the subsurface reservoirs can provide a steady water source. This can be increased in periods when total failure of the surface water resources occurs. Moreover, in contrast to the negative attitude towards groundwater development, it is believed that a positive effect on the environment can be achieved if enough wells are drilled and appropriately equipped in order to avoid the over concentration of animals in one area.

A positive environmental impact is believed to take place because of the availability of an assured water source which reduces the range of movement of nomads, as well as the encroachment of nomads into these areas where water resources have remained available. In many cases, these have been the same areas where wildlife has also taken refuge. The competition between nomads and their herds and wildlife feeding on the same resources has ended with a loss in the population of wildlife.

The gradual stabilization of the nomadic populations, although believed by some conservative sociologists to be a negative development, seems to us to be unavoidable in face of the changes in the socio-political changes in the African countries. On the one hand, these changes have limited the range of possibilities of migration due to the artificial limits imposed by the political borders. On the other hand, the marginal regions which were the traditional habitat of the nomads have become a prospective outlet for the explosion of the population in regions of higher potential. For example, in the Garissa district, along the Tana River, the District Commissioner and officers have reported that many people have started to settle along the river, pumping its water in order to grow various cash crops. Once this way of settlement becomes more popular, the growth rate of the population along the river may grow exponentially, causing destruction of the riverine forest. This will cause erosion of its embankment and an encroachment of desertification to all the land bordering the river. This will deprive the nomadic population of the flood irrigated lands along the river, not to speak of the wildlife which found refuge in the riverine forest and the surrounding flood irrigated lands.

The settlements along the river banks are to be carried out in the framework of a regional master plan and will be located beyond the belt of the riverine forest. The water supply for livestock will be provided by a network of wells spread out according to the carrying capacity of the area. Such a project will induce the nomads, depending on the flood plains, to settle down without causing damage to the natural environment and will encourage them to become part of the settlement process of these regions, envisaged by the population explosion in neighboring regions.

It is recommended to develop the surface and subsurface water resources in a gradual way, namely starting with small, simple pumping devices and to advance, stage by stage, to more complicated means. It is believed that the need to handle and maintain this machinery will bring about the development of local know-how, as happened when the motorized car was introduced in these regions.

Another important aspect in environmental protection will be the economic security which the nomad population will achieve. This will encourage them to abandon traditional ways of life, which has a negative impact on the environment, such as communal ownership of land. It should be emphasized again, that any settlement project will have to be carried out in the framework of a comprehensive development project which will cover all aspects of the environment, including the safeguarding of regions set aside for natural reserves. It is expected that the increase in the food and water supply will cause an increase in the number of wildlife around settlements, similar to what has happened in the Negev desert of Israel.

A.1.6 Education

A major obstruction to the advancement of the region is the reluctance of educated people, like high education teachers, agronomists, instructors and technicians to settle in this area. This is due to the distance from an advanced urban center and thus the lack of an educational and social infrastructure. Experience from The J. Blaustein Institute of Desert Research in Israel has shown that the success of such an institute, situated far away from an urban center, can be achieved when adequate facilities for pre-school, primary and secondary education, as well as technicians' aid facilities are guaranteed for an adequate standard of living. It is thus suggested to initiate the building of a regional education center, starting from primary school to university in Wajir, the administration center of the province. This center will serve the whole region and will undoubtedly also boost the development of this town. The scientists and technicians will be able to commute from this center to the research and field stations, as well as to other projects in the neighboring districts. It is advised that a detailed scheme for such a center be prepared by the Kenyan government to be presented to a donor agency.

An agriculture school and research center will be part of this center. It is suggested to put special emphasize at the research center on irrigated pasture and fodder crops, as well as dairy cows on a zero grazing basis. The research will also include various types of irrigated agriculture.

A.1.7 The Construction of Dams

As was explained in the previous chapters, the philosophy intrinsic in the plan of Progressive Development suggests the gradual build up from low investment simple projects to more expensive, complex and larger scale projects, which require not only large investments, but a permanent complex infrastructure and thus a special administration for control and maintenance. Thus the shallow and later the deep groundwater are the first targets of development. Yet, in the future when the development of irrigated agriculture and urban demand reaches the level that the groundwater resources become exhausted, then the stage of building of dams and projects of water transfer by canals or pipes from one part of the province, or even beyond, can start.

A few dams were already built in this and in neighboring regions, but their affect on the economy and stable water supply to the people was below expectation.

In the report submitted to the Kenyan government a list of such projects has been given and it is beyond the scope of the present book to go into the details of this issue. Yet it should be emphasized that the planning of a dam, or the extension of an irrigation system of an existing dam, should be part of a general plan of irrigation backed by an extension service. Special emphasize will be put on the introduction of irrigation methods which will avoid the waste of water and thus soil salination processes. A demonstration farm must be included in each such plan, in order to guarantee the efficient utilization of the vast investment involved in building a dam and a water distribution system.

Appendix B
A Call for Global Action Replant the Dry Lands!

A.S. Issar, M.L. Arias, P. Berliner, J. Bogardi, H. Faure, A.M. Lezine, D. Prinz, P. Rognon, A. Riedaker, T. Tanaka, N. Tsiourtis, T.S. Liu, and J. Williams

Make better use of local water resources, provide food, wood and land for the people and mitigate the global greenhouse effect

B.1 Executive Summary

The following "CALL FOR GLOBAL ACTION" summarizes the discussions carried out by a group of scientists who convened at the Division of Water Sciences of UNESCO, in Paris, on the 14th and 15th December 1998. The group was concerned with the forecasts about the negative impacts of the greenhouse effect on the hydrological cycle in general, and on that of many of the dry zones of the world in particular, as these impacts are disastrous for the vulnerable socio-economic systems which subsist in these regions.

The group, looking for ways to try and mitigate these negative impacts, recommends that in addition to the targets agreed upon in Kyoto – namely, cutting down of carbon emissions and subsidizing sequestration of atmospheric carbon by forestation in the humid regions – an effort, on the international scale, should also be made to encourage planting of trees in the dry lands. Such activity should be considered, despite the natural constraint with regard to the limited precipitation of these regions, which may cause such projects to be regarded as uneconomical, when only carbon sequestration is considered. It is claimed that the additional benefits to the developing countries from such projects, by providing new food and wood resources as well as combating desertification, justifies the provision of extra support for such actions. To this end, the committee recommends that the international team working out the rules for subsidizing carbon sequestration projects should procure a special

A.S. Issar (✉)
Ben Gurion University of the Negev, J. Blaustein Institutes for Desert Research,
Sede Boker Campus 84990, Israel
e-mail: issar@bgu.ac.il

By the members of the steering committee for the project envisioned in the framework of World Water Vision and The International Hydrological Program (IHP), UNESCO.

financial mechanism for the countries in the dry lands, namely, countries classified as those combatting desertification.

If such subsidies are made available, then the constraints of limited precipitation can be overcome, in the first place, by planting trees for food and forage, irrigated through the efficient use of the existing surface and groundwater resources existing in these regions. That such resources, which are currently going to waste, do exist, has been proven by hydrological and agro-forestation research projects, which have been carried out in these regions.

Further similar projects, to be carried out following investigations with regard to their environmental sustainability, will have a positive impact on the natural environment, by reducing erosion of soils and checking encroachment of sand dunes. In many of these countries, such projects will involve the rejuvenation of ancient agricultural systems, which were destroyed due to climate changes and human incompetence. Modern, yet sustainable, agricultural methods will help in promoting the economy of these countries, and, at the same time will have a positive impact on the global environment. In other regions, only marginally economical agricultural practices, such as grazing, can be replaced by the replanting of the original savanna vegetation to create natural park lands, animate bio-diversity and promote the tourist industry.

The committees in charge of the accomplishment and realization of the Kyoto and Buenos Aires protocols, as well as the international society of scientists, international governmental and NGO agencies, are herewith requested to put on their agenda this call for action, secure the funds and initiate projects to make better use of the water resources existing in the dry lands, in order to turn them into greener areas, and in this way, combat desertification, assist the economy of the developing countries and in parallel, mitigate the greenhouse effect.

B.2 Introduction

Many observations indicate an increase in the concentration of carbon dioxide in the atmosphere and thus, climatic warming, as a greenhouse effect is the most logical conclusion of this condition. In order to reduce the possible environmental impacts of such an increase, an international agreement was signed by the representatives of the countries participating in the Kyoto conference, to curb the amount of emissions by the industrialized countries.

The goal is to reach a reduction of 5% in the period of 2008–2012, from the level of emissions during 1990. In addition to the reduction effort, the Kyoto protocol (article 3.3) recognizes that projects, which involve sequestration of atmospheric carbon could be used to offset emissions. Carbon sinks can be created by afforestation projects, due to the fact that trees and other vegetation absorb carbon dioxide by photosynthesis and fix it as cellulose. Thus, it is expected that an international effort of rejuvenation and creation of new carbon sinks, through re- and afforestation projects, will develop. This will be stimulated by a further economic payoff, in addition to the usual profit from forestry projects in humid regions. Subsequent to the signing of the Protocol, a certain sum will be paid per ton of sequestered carbon. Negotiations are still underway with regard to the rules that will decide the accreditation of sequestration, namely which activities and what time dimensions will be taken into account. Once the Kyoto protocol's rules encouraging the reward for the offset of emissions are formalized, it is expected that many countries will be encouraged to proceed with large scale afforestation projects as, for example, in Western Australia where "It is estimated that it will be possible to sustain a carbon sink of approximately 200 million tonnes assuming a 30-year accounting cycle" [1].

Yet, based on past experience, it can be forecast that the major effort of re- and afforestation, on the international level, will be invested in the humid parts of the world, since the growth rate and thus, the sequestration rate, as well as cost/benefit ratios from wood products, exceed those of the dry lands. It is estimated that areas with between 400–600 mm of precipitation per year sequester annually between 103–140 tonnes carbon per ha, while areas over which annual precipitation is only 250–400 mm only sequester an average of 30 tonnes, and those over which annual precipitation is less than 250 mm, sequester only 5 tonnes of carbon per ha [1]. Nevertheless, it should be taken into consideration that in many humid countries, re-afforestation projects on a wide scale, have their socio-economic constraints, due to the competing demands for land for many other purposes, such as urbanization, agriculture, industry, transportation facilities etc. Thus, the gain from afforestation projects, have to counterbalance (at least) the value of land, and loss of alternative income, arising from alternative usage.

This is not the case in most dry lands, where the density of the population is low, and thus the demand for land is insubstantial. Surveying the Earth from a satellite vantage point reveals that the yellowish color, demarcating semiarid and arid zones, covers about 30% of the continents. In these regions, population and agriculture are concentrated in areas where water is obtainable, either from rivers or shallow groundwater. When one looks beyond these areas, the densities of population and vegetation diminish abruptly and vast stretches of land, in many cases covered by sand dunes or loess deposits, are practically treeless. If only a part of these vast stretches of land can be forested, then large amounts of carbon dioxide can be sequestered. This is, of course, more easily wished than done, as the main reason such land is underutilised and bare, is the scarcity of water – the essential requirement for the existence of bio and phyto systems. This awareness is deeply imprinted in human thinking, which may explain the fact that to date, practically no serious attempt has been made to suggest promotion of carbon sinks, in the form of trees, in the dry lands. One can learn of this failure while surfing the Internet for afforestation projects aimed at fixing carbon dioxide. Hardly any of the voluminous references, dealing with this issue, relate to the yellow belts of our globe.

Thus the first and main question which must be answered before one recommends the planting of trees in these regions, is whether enough water resources can be procured to enable the promotion of woodlands. Once this question is answered in the positive, the next question must be whether such an initiative is practicable from the economic point of view?

B.3 On the Availability of Untapped Water Resources in the Dry Lands

It is the opinion of the authors of the present document that this questions can be answered in the positive, and that the main restraint is human inventiveness. This is because hydrological and hydrogeological investigations carried out in the arid and semi-arid regions all over the world show that ample resources of water can still be found in these regions, and as in the past, it will be a question of human ingenuity whether this water can be utilized to a positive end. Indeed, since the invention of agriculture about ten thousand years ago, farming societies have found ways to make use of the available water resources in midst of the desert.

If man could achieve this when technology was rudimentary, we should be able to make use of the existing water resources, even if they are scarcer, due to the negative impact of climate change in modern times. Moreover, in many areas forsaken by past civilizations unable to cope with the worsening environmental and economic conditions, their terraced lands now abandoned, water is now wasted, flowing freely and eroding the soils, year by year.

Once it is decided to reforest these regions, the water available for such projects can come from the following sources:

1. Surface water. Various investigations of the hydrological regime in arid countries (Reviewed in Issar and Resnick [2], and in Prinz [3]) have shown that the percentage of the volume of water of runoff, relative to the total volume of the precipitation over a certain area, decreases as the size of the catchment area increases. This means that the larger the catchment basin, the more water is lost by evaporation. The ancient farmers were aware of this, and diverted water from small catchment areas to irrigate their terraces, where they grew fruit trees and cereals. These methods and even the old abandoned structures built for "water harvesting" in many places, can be renovated, and the water which is now wasted can be utilized for reforestation.

2. Groundwater. Based on the results of the many surveys (extended list of references in Lerner et al. [4]) it can be guaranteed that under most of the deserts of the world there exist aquifers containing non-utilized fresh and/or brackish groundwater. Some of these aquifers are recharged yearly and some only once in a few years by precipitation falling on the desert, by infiltration from floods into alluvial fans, or by subsurface flow from adjacent highlands areas where precipitation is more abundant. In many areas, this subsurface water emerges at the surface and evaporates to form salt marshes.

Of special interest is the subsurface water found in sand dunes, even in areas receiving not more than 80 mm of rain per year. It was found (extended list of references in Adar et al. [5]) that in such areas, the moisture content of the soil at the end of the dry season under forested dunes, is higher than that of bare dunes, which emphasizes the positive impact of tree cover on the subsurface water balance of the dunes.

3. Paleo-groundwater. Under most of the deserts of the world, there exist tremendous amounts of groundwater which were recharged when climate conditions were much more favorable, tens of thousands of years ago (extended list of references in Issar and Nativ [6]). This water moves very slowly, either towards springs (like the Mound-Springs in Central Australia, salt marshes (like the schotts of the Sahara) or towards inland lakes (like the Dead Sea). In Australia, Saudi Arabia, Libya, Jordan and Israel, some of this water is utilized for agricultural and industrial purposes. Needless to say, the use of this water means its "mining", which may be regarded as a non-sustainable approach, yet at the same time, one should also regard this resource as a potential to ameliorate the ill effects of the burning of fossil fuels. Thus, if humanity cannot avoid the use of fossil fuels, and thus causes harm to environment, why should not fossil – namely paleo – water be used to correct this harm? Of course, once other economical fuel sources can be developed that are more friendly to the environment and still supply the energy required by human society, then the use of this water can also be discontinued.

4. Sewage from urban centers. Due to the current world urbanization trend, metropolitan centers in the arid and semi-arid parts of the world and in the more humid parts of the world are growing continuously. These centers produce large amounts of sewage which has to be treated and then disposed of. Utilizing this source for plantations has its environmental and economic benefits as well as constraints, which must be considered.

B.4 Aspects of Sustainability

During historical and pre-historical periods, vast areas, which are presently bare dry lands, sustained an arboreal vegetation. This means that, from the point of view of the ecological system, once the original species are reintroduced, (and this can be done on the basis of the pollen assemblage found buried in the subsurface) reforestation can be a rejuvenation process, rather than the invasion of an alien system. Yet, the knowledge acquired recently that the more green conditions of the past were a function of more humid climate conditions, requires us to put special emphasis on maintaining the sustainablity of the hydrological system from the point of view of the quantity, as well as the quality of the water. In other words, measures have to be taken to manage the resources so to avoid its deterioration, as for example, by salt concentration in the soil and the subsurface. Such measures include water harvesting methods by terracing, which introduce extra water, otherwise wasted, into the hydrological cycle. This enables the management of the water, soil and tree system on a higher level of efficiency and supply.

To ensure sustainability, projects should be planned and carried out as an integral component of a comprehensive plan of regional development, preceded by hydrological, pedological and ecological research, and followed up by monitoring and re-assessment procedures.

B.5 Economic Aspects

It would be difficult to dispute the argument that if the forestation of dry lands had been feasible from the economic point of view, then these areas would have been green by now. Yet, in the same time, the recent negative history of the rain forest shows that what pushes many activities of human society are the short term gains, rather than the long term benefits, not to speak of the rather low value assigned to benefits, like bio-diversity, that can not be quantified.

This being said, it is here argued that plantations in the dry lands can still become profitable enterprises, due to the incorporation of immediate local agricultural benefits, long-term local fringe benefits, and long-term global benefits. However, to encourage the spin off of such enterprises, enlarged subsidies for sequestration of carbon dioxide are required.

The committee was encouraged by the account, given by the chairman, of a plantation project of thousands of hectares of olive trees carried out by a private enterprise in the highlands desert of Southern Israel (average annual precipitation 100 mm), to be irrigated by brackish (ca. 4,000 mg/l TDS) paleo-water.

Needless to say, an economic feasibility investigation of a specific environmental sustainability project must be undertaken before any such project is launched, as was the case with the above mentioned project. What can be said at the present stage, is that plantations, irrigated by flood water and by groundwater, extend over thousands of hectares, all over the dry lands of the world, and their further expansion, encouraged by special subsidies will most probably contribute, not only to the sequestering of carbon dioxide, but also to the welfare of the population of these countries. In a world where water resources are scarce and the supply of food is becoming increasingly uncertain, the development of these additional sources of water, food and wood is a challenge not to be neglected.

B.6 Conclusion, Recommendations and a Few Words of Caution

1. The present call for action is aimed at the agencies in charge of carrying out the recommendations of the Kyoto protocol, to promote global comprehension and practical steps with regard to the role which large-scale plantations in the dry lands may play in sequestering atmospheric carbon. For this purpose, the committee recommends special subsidies in order to encourage such plantations in these regions. This, in the framework of the steps envisioned to carry out the recommendations of the Kyoto protocol in order to offset carbon emissions.
2. This call is also addressed to the international scientific community, who was first to ring the alarm, with regard to the negative impacts of the greenhouse effect. The steering committee suggests the promotion of an international web

of scientific centers, interested to be involved in research projects which will advance knowledge concerning all aspects of reforestation of dry lands. This web will enable the formation of database terminals, at once, and a net-work of data exchange, and initiation of cooperate projects. Cooperation between scientists in industrialized and developing dry countries will be one of the main goals of this net-work.

3. The steering committee recommends to convene in the year 2010 an international conference, which will deal with the various scientific aspects of plantations in the dry lands, namely the climatic, the hydrological, the ecological and the economical aspects.
4. A call is addressed to the other UN agencies, EU authorities, USA government offices and NGO foundations, to commit the required budgets in order to promote research, pilot projects, and later on full fledged projects of reforesting the dry lands. The initiation of Bi & multi-lateral projects, involving developed and developing countries is highly recommended.
5. The international media, whether scientific journals, popular science journals, or the general press, which have done so much to make the people of the world aware of the looming hazards of Global Change are requested to aid in promoting the idea of planting the dry lands, yet, without subduing the demand from the industrial countries for the reduction of emissions required by the Kyoto protocol,
6. Once the understanding of the direct and fringe benefits, which such a project could bring forth are engrossed by the government agencies, scientists, NGOs, and the media, an action on a global scale have good chances to follow. When this happens, planners, entrepreneurs and executives, must be aware of the fact that such projects must be approached in a comprehensive way, in order to ensure the sustainability of the environmental system. Thus any scheme must be based on the results of a comprehensive research and development plan, which will involve forest experts, hydrologists, biologists, soil scientists, economists etc. Once this approach is adopted and a monitoring system is installed to warn against irreversible results, then there is hope for this project to spread over the more arid parts of the globe and benefit human society and global natural environment.

References

1. S. Shea, G. Butcher, P. Bartle Ritson, P. Biggs, Western Australia's Development and Future Prospects for Tree Crop Industries, Australia's Paper and Forestry Forum, 18–19 May 1998, Department of Conservation and land Management, p. 18 (1998).
2. A. S. Issar and S. D. Resnick, *Runoff, Infiltration and Subsurface Flow of Water in Arid and Semi-Arid Regions*, Kluwer Academic Publishers, Dordrecht, The Netherlands (1996).
3. D. Prinz, Water harvesting for afforestation in dry areas, Papers presented at the meeting of the steering committee for the project Replant the Dry lands, UNESCO, 14–15 December (1998).
4. D. N. Lerner, A. S. Issar, I. Simmers, *Groundwater Recharge: A Guide to Understanding and Estimating Natural Recharge*, A. A. Balkema Publishers, The Netherlands and IAH (1990).

5. E. M. Adar, I. Gev, P. Berliner, and A. S. Issar. The effect of forestation on a shallow groundwater reservoir in an arid sand dune terrain. *Journal of Arid Land Studies*, 58, 259–262 (1995).
6. A. S. Issar and R. Nativ, Water beneath deserts: Keys to the past, a resource for the present. *Episodes*, 11(4), 256–262 (1988).

Lightning Source UK Ltd.
Milton Keynes UK
27 October 2010

161933UK00005B/157/P

9 783642 106392